P9-DTT-514

LOOKING UP

LOOKING UP

THE TRUE ADVENTURES OF A STORM-CHASING WEATHER NERD

MATTHEW CAPPUCCI

PEGASUS BOOKS
NEW YORK LONDON

LOOKING UP

Pegasus Books, Ltd.
148 West 37th Street, 13th Floor
New York, NY 10018

First Pegasus Books cloth edition August 2022

Interior design by Maria Fernandez

ISBN: 978-1-63936-201-1

10 9 8 7 6 5 4 3 2 1

Printed in the United States of America
Distributed by Simon & Schuster
www.pegasusbooks.com

To my parents—who got handed a quirky kid and

gave the world to him every step of the way,

even if that meant following a limousine for 40 miles

on the highway so five-year-old him could gawk at it,

buying $50 in pennies at the bank

so seven-year-old him could sort through them,

sitting at a railroad crossing for an eternity waiting for

a train to activate the gates and flashers, or

driving ten-year-old him to the beach at 2 A.M.

to watch a thunderstorm over the water.

I also promised my mother that I'd thank her in my speech when I received my first Emmy award. That hasn't happened yet, so, Mom, a book dedication will have to do for now. We'll get the Emmy soon enough.

I should probably thank my sister, too. Though she's four years younger than me, I'm still (slightly) afraid of her. If nothing else, she kept me humble.

Also dedicated to Waffle House. Sponsor me already damnit.

The Beginning

If you're ever looking to get over a fear of flying, spending ten hours on a sweat-scented double-decker bus from Boston to Washington, DC, will do the trick. It was July 2012, and while every other friend of mine was spending the summer caddying or refereeing youth soccer games, I was on my way to weather camp. Apparently, it's a real thing.

I was fourteen years old and had just completed my freshman year of high school on Cape Cod in Massachusetts. It wasn't the first time my activities didn't exactly line up with those of my peers.

Most elementary school kids spent recess trading Yu-Gi-Oh! cards or swapping Silly Bands; I was staring at the sky and journaling. My best friends in second grade were my teacher and the ladies in the front office. (Mrs. Monska, the head secretary, had a Jelly Belly machine on her desk. I visited every morning without fail.)

My classmates all had PlayStations, but I spent my First Communion money on a bulky camcorder. Any distant rumble of thunder sent me running to the garage to grab my bike for an impromptu "storm chase" around the neighborhood cul-de-sac. The footage I captured fell tragically short of *National Geographic* quality, but it was a start.

"Dad, I got another lightning strike," I can be heard exclaiming in lisp-twinged second grade glee, my clunky camera work entirely missing the electrical discharge in question. "It was a pink one." I then proceed to explain to the "folks" watching how ribbon lightning works, what a wall cloud is, and why the clouds are so dark. Despite growing up in the Boston suburbs, I sounded like a veteran Great Plains broadcaster. Since then, my passion has never wavered nor left me. On the contrary, it matured as I did.

I gave daily weather reports at school to my fourth grade class during Ms. DeLorenzo's morning meeting. My field day and recess forecasts blared over the Indian Brook Elementary School intercom. My parents grew accustomed to 2:00 A.M. taps on the shoulder, nine-year-old me whispering to rouse them from their slumber. "Can we go to the beach?" I would beg, desperate for a view of whatever was happening in the sky.

Every year on my birthday, I corralled my family into the front yard to lay on a blanket and stare skyward as the Perseid meteor shower slung shooting stars across the heavens. And if thundersnow ever entered the forecast, I could go days on end without sleep.

Middle school passed similarly. The other kids took to sports and playing *Call of Duty*. I memorized the periodic table and landscaped in the summers to earn money for a laptop. In seventh grade, I enrolled in a class called Helping Others While Learning, taught by Warren Phillips, a renowned science educator who was inducted into the National Teachers Hall of Fame that year. It proved a pivotal turning point.

Mr. Phillips combined service learning with science geekery to constantly push the limits of what was possible in a classroom. He led each of us through making a gel electrophoresis, implemented an ambitious school-wide recycling project and, most importantly, was awarded a grant that allowed for the production of a student newscast. Three seconds into our first filming, I was hooked.

As it turns out, middle school isn't built for quirky kids who aim to grow up and become scientists. I stuck out like a sore thumb. I was adored by administrators and teachers—especially within the Science Department—but I certainly didn't blend in with my peers. Most were too wrapped up in drama and *Doodle Jump*. As the midpoint of eighth grade approached, I was looking for an escape plan.

That turned out to be Sturgis, a charter high school located in Hyannis, Massachusetts, about thirty minutes away from my hometown of Plymouth (yes, where the Pilgrims landed). Affectionately referred to as the "island of misfits," Sturgis had a reputation as being a place where just about anyone could fit in (maybe even me). And the academics were second to none.

Erected in a converted furniture store, Sturgis was held together with duct tape and bubble gum. The roof leaked whenever it rained or snowed; we had no cafeteria or gym, and the art room was a house the school had purchased on an adjacent street. The campus wasn't much to look at. Somebody once left bananas in a locker during winter break, resulting in a fruit fly infestation we discovered in January. (The school decided to combat the pests by leaving dishes of apple cider vinegar all over, which promptly spilled and left sticky floors for more than a year.)

Because the school was located next to a row of shops, restaurants, and art galleries, it wasn't unusual for tourists to stroll through the building on a whim. Sometimes people walked their dogs through the main hallway. The architecture was labyrinthine.

The teachers were equally eccentric. My history teacher had a "pilgrim voice" and character he occasionally slipped into; another routinely donned a colander to impersonate the former Soviet satellite "Sputnik." My Spanish teacher had just turned twenty-two, and my mustache-sporting art teacher sprinted into class the first day wielding a hammer and an unhinged stapler.

Our chemistry teacher sometimes taught while doing snow angels on the floor "for [his] bad back." Mr. Carah, the physics teacher, could get away with saying just about anything—no joke was too off the wall. And my math teacher once sprinted home from class mid-lecture to make sure he had shut off his stove.

Since the school didn't have a designated vehicle to transport goods, they relied on a donated vehicle that the administration fondly referred to as "the creeper van" to shuttle around band and student council equipment. Every March, we had a fundraiser called Tape a Teacher, where a $5 bill would earn students a piece of duct tape to affix a teacher to the wall. One year we inadvertently taped Mr. Dunigan-AtLee, a math teacher, to a utility pole in front of the school with his arms outstretched in a Christ-like state . . . on what turned out to be Good Friday (that led to an unfortunate article in the *Cape Cod Times*). We also made the news when four hundred copies of a yearbook reading STURGIS CHARTER *pubic* SCHOOL were delivered. That may or may not have been my fault.

It was an erratic, unconventional school where everyone was as quirky as I. Within a week, I knew I was home.

Now, freshman year at Sturgis had come to a close, and I was on my way to the nation's capital for a two-week weather camp at Howard University. Ever the tireless parent, my mother had finally "found a place for me" after scouring the internet for opportunities. The National Oceanic and Atmospheric Administration–funded summer program featured an immersive curriculum under the auspices of a former National Weather Service meteorologist.

More importantly, it offered the prospect of finding myself among a dozen other weather-obsessed nerds like me. For the first time in my life, my unshakeable fixation with weather would be the norm. I knew I had to apply. Weeks later, I kissed the jam-packed oversized manila envelope good-bye, crossed my fingers, and waited. After a month of checking the mailbox daily, good news arrived—the only catch was that we had to get ourselves to Washington, DC.

"Try the Megabus!" my high school bus driver said when I told her the news. (Since Sturgis was a charter school, we all had to rent our own bus and hire a driver.) Her name was George, and raspy shouts of road rage–induced profanity were a staple of our afternoon commutes. If she wasn't texting or scrolling through the TD Bank app while driving, she was dispensing somewhat questionable life advice. But when she mentioned $5 bus fares to Washington, I paid attention.

It would be my first time away from home, and I was a bit apprehensive. My saintly mother agreed to accompany me to Washington, seemingly enthused at the prospect of a bus trip down the Northeast Corridor. Our mutual excitement over a bargain prevailed as we headed to South Station in Boston one morning in early July. We had no idea what we were getting ourselves into.

The hulking double-decker bus lumbered into the station ninety minutes late, reeking of diesel fumes and whatever's inside blue portable toilet water. Hordes of passengers rushed aboard in a stampede, haphazardly chucking luggage at the driver for stowage in the under-cab compartments. My mother and I climbed to the second floor of the top-heavy bus. (We wound up shoehorned in the back, but, to my relief, near the stairs—Megabus drivers had a history of plowing into shallow overpasses.)

Saying it was a harrowing journey would be an understatement. The journey quickly devolved into a hodgepodge of cascading headaches, but somehow it all added to the excitement of my first weather camp.

In Union, Connecticut, we had to backtrack a half hour after leaving a woman behind during a pit stop at Burger King. New York City traffic snarled us with ninety-minute delay, and the first floor toilet overflowed into the cabin in Newark. In Hoboken, New Jersey, a hitchhiker attempted to board the bus; shortly thereafter, a fistfight almost broke out between two women over a pair of stolen sunglasses. The woman to my left had found it fitting to frame me.

Temperatures inside the bus rose to 109.2 degrees in Philadelphia thanks to air conditioning issues, which I know because I was carrying a pocket thermometer. By the time we got to Washington amid the unbearable heatwave, I felt famished, lightheaded, and dizzy.

Once we arrived in Washington, I transitioned to just plain terrified. I was four hundred miles from home, had just been granted my first cell phone, and was being dropped in an unfamiliar location with unfamiliar people. Reality hit fast. After a dinner of chicken tenders and soft pretzels, I fell into a fitful sleep, anxiously dreading what lay ahead.

The next morning I stared emptily at my breakfast plate, too strung up to even eat a slice of bacon (anyone who knows me will immediately realize how significant that is). Around 10:00 A.M., my mother and I hopped on the Metro and rode to Howard University, where I planned to bid her farewell. I wasn't concerned about being away from my parents, but spending two weeks in close quarters with eleven peers would be daunting.

I wasn't feeling much better after the program director's introductory remarks. Then came time for the camp's leader, meteorologist Mike Mogil, to speak. About a sentence in, I yelped, leaping up as though my pants were on fire. Something in my pocket was buzzing and squealing. Everyone turned to face me.

I suddenly remembered I was carrying my travel-sized NOAA weather alert radio. I blushed, wishing I could melt into the floor. I hastily mumbled an apology, trying to silence the still-screeching radio.

"Wait a second," Mr. Mogil said. "Are we under some type of alert?" He extended his hand, motioning for the radio.

"The National Weather Service has issued Severe Thunderstorm Watch 413 in effect until 9:00 P.M. tonight," barked the automated Perfect Paul voice. The room erupted into shouts and cheers; Mr. Mogil cracked a grin.

"Welcome to weather camp!" he bellowed. In an instant, I realized I was in my natural habitat.

The kids alongside me were every bit as eager about the stormy forecast as I was. Hours later, I found myself analyzing the 3:00 P.M. data, half a dozen other campers surrounding me as I explained the incoming charts and figures splayed across my laptop. It kicked off the most amazing two weeks of my life. My lifelong obsession with weather made me the star of the group and earned me Mr. Mogil's invitation to the American Meteorological Society's annual broadcast meteorology conference in Boston later that summer. The following year, I vowed to not only attend the conference, but to present at it.

Having seen the presentations delivered from meteorologists of every background, I knew my research was up to snuff. During my sophomore year of high school, I compiled my work into an abstract titled "Outflow Boundary-Related Waterspouts: Advanced Detection and Warning." In January 2013, I submitted it to the American Meteorological Society for consideration. By March, I was booking a ticket to Nashville.

Being fifteen years old, however, meant that I wasn't permitted to rent a hotel, procure a rental car, or even fly alone. I knew I would be able to, but society didn't agree. Thankfully, my mother agreed to be my chaperone.

Because the hotel was exclusively occupied by meteorologists, every elevator ride, momentary instance of hallway eye contact, or breakfast line turned into a speed round of networking. But no one wanted to talk to me; after all, how could I be a meteorologist? They defaulted instead to my mother, a longtime pediatric nurse at Boston Children's Hospital. She could sew Humpty Dumpty back together again and save his life after a great fall, but clouds and water vapor were *my* thing. I felt like a shadow.

I drifted quietly around the conference for two days and, save for the occasional pleasantry offered by an observant and sympathetic small-town meteorologist, I was given the cold shoulder. Then came Wednesday. I silently sat in the back, reviewing my slides, making sure I knew where to pepper in my occasional humorous quips and moments of levity. As the presenter before me concluded her remarks, I began making my way to the lectern.

"Next up, we have Matthew Cappucci, who will be speaking to us regarding outflow boundary-related waterspouts," the presenter announced. In the online application, I had selected the university/graduate student box, as that was the closest option to tenth grader. But they accepted my research anyway.

Now, I stood at the podium a thousand miles from home, explaining my hypothesis, and why I believed a series of weak, erratic, and quick-hitting tornadoes had made appearances along the Massachusetts coastline during the summer of 2012. Not a peep could be heard in the audience. I zeroed in on three case studies I had prepared, each featuring documented examples of waterspouts, or tornadoes over water, sweeping ashore and causing damage. They all formed on days when storms were outflow dominant, exhaling more air than they were ingesting. I wove together my theory, explaining why low-level lapse rates, vorticity stretching, and downdraft surges were vital. I supplemented my claims with meteorological observations.

In essence, I asserted that the cool breeze rushing away from an approaching thunderstorm could occasionally spin up invisible horizontal tubes of air a thousand or so feet above the ground. Ordinarily, they aren't problematic. But when that tube intersects the coastline at an oblique angle, it can become fragmented—a series of smaller horizontal vortices resulting. If the temperature contrast between the air within the storm and that ahead of it is great enough, warm air forced upward ahead of the gust front can tilt a tube, stretch it vertically, and form a waterspout. Under the influence of capriciously moving cold air from behind, those spouts can easily be steered toward land.

It had been enough to convince the local National Weather Service office to append additional statements at the end of their severe weather warnings, the bulletins first utilized in September 2013. The crowd nodded in agreement, exchanging occasional glances with one another. I heard a few murmurs. In that instant, the age difference between the audience and myself evaporated. It dawned on me that the group no longer considered me "the kid," but rather a budding colleague. Someone who shared the same passion as everyone else in the room. At least that's what I told myself.

As I wrapped up my speech, cognizant of the blinking yellow timer, I opened up for questions. An older gentleman, who from a distance appeared to be in his upper fifties or lower sixties, sauntered up to the aisle microphone. I awaited the question, ejecting my flash drive and fidgeting with my clicker.

"First off, you can have my job in far less than six years," the man said. Without looking up from my podium-mounted monitor, I recognized the voice instantly. My eyes widened. "The second thing I want to say is that, at your age, at this point in your life, to be able to put this together, break new ground and deliver it is absolutely unbelievable," he continued. It was Harvey Leonard, chief meteorologist at WCVB-TV in Boston. I had grown up watching him religiously. I had a saying in my household: "When Harvey's talking, you're not." He was my meteorological idol. Behind the podium, my knees shook. (He's been a friend and a mentor ever since.)

At the coffee break following my talk, I experienced something I had never encountered before: I was popular. Meteorologists congratulated me, joked about buying me a beer, and even approached me with questions or observations of their own. Now a degreed meteorologist, my colleagues have become some of my closest friends. It took years of wandering around like a tumbleweed, staring up at the sky, but I've finally found my people. Nowadays, seven or eight conferences later, I'm finally old enough to get that beer.

Suddenly being Mr. Popular was new to me. The conference lasted another two days, and everyone wanted to talk to me. I was all for it. It was as though each of the two hundred meteorologists there saw a piece of themselves in me, remembering back to when they were in my shoes. I frequently think back to the kindness they extended, hoping to use the platform I have today to pay it forward.

As I headed back to Cape Cod for the start of junior year at Sturgis, I was busy hatching a plan on how I could capitalize on my newfound momentum. The conference had reignited a spark beneath me, and I wasn't about to let that go to waste. I continued cataloguing weather events and decided to try my hand at writing for local newspapers.

On October 29, 2012, the fringe effects of Hurricane Sandy, which had ravaged New Jersey the day before, arrived in the Northeast, where they knocked out power and snapped tree branches all across Plymouth and Cape Cod. I wrote eight hundred words and emailed it to the local newspaper to see if they were interested. They ran my piece on "The Science of Sandy" in their next biweekly edition.

That was my first time getting to actively share my enthusiasm for all things atmospheric with the public. Days later, a 100 mph microburst, or downward rush of severe winds from a thunderstorm, caused serious damage and sank several boats two towns over from me. Once again, I contributed a write-up.

I was granted a regular column in Plymouth's *Old Colony Memorial*, where I quickly began submitting thousand-word explainers deconstructing local, regional, and national meteorological happenings. I knew my minuscule audience was primarily comprised of nursing home residents and retirees, but it was something. I hoped someone—anyone—was learning something new.

Apparently, they were. Soon thereafter, I was treated to a nice little surprise. I opened up that week's edition of the small-town paper, immediately finding my name. But it wasn't in an article I had written. It was a letter to the editor that read "Hats off to Matthew Cappucci." I quickly snatched the page for a closer look.

The paragraph-long submission was from a gentleman who apparently enjoyed a piece of mine on hurricanes. I was ecstatic. I had a living, breathing reader! The letter was signed by Eric J. Heller, a local resident who lived a few miles north. After asking the editor to connect us via email, I sent him a note of appreciation for taking the time to pen such kind words. We shared a brief online exchange, I thanked him, and then returned to an otherwise routine world of homework and weather maps. I was fifteen. It seemed insignificant.

It turns out I was wrong.

Senior Year

"Who are you!?" I whooped into the phone. It was November of my senior year, and Thomas, the homeschool student I was tutoring, was shrieking with laughter. "What do you want!? Why do you keep calling me!? I don't *want* your solar panels!"

I belligerently hung up, tossing my flip phone onto the table before cracking a smile and shaking my head. Thomas, who had been working to stifle his giggles, erupted into a fit of guffaws.

"And that, Thomas, is how you take care of telemarketers," I said professorially. He was practically snorting with merriment. My phone had rung four times during my one-hour Latin tutoring session with Thomas, who was thirteen years old and lived at home with his overbearing and deeply religious guardians. I was the only outside interaction he was permitted to have. Phone spammers were certainly not going to interrupt me at work.

Thomas's adoptive parents, his aunt and uncle, were overwhelmed with life. His uncle Richard was a former Green Beret who suffered from post-traumatic stress disorder. He isolated himself in the house, shades drawn, with air purifiers in every room. Thomas's aunt Debbie was a kindly woman who awoke early every morning to pray and care for Thomas's sister, Veronica, who had profound handicaps and was nonverbal.

Thomas was brilliant and an erudite learner, but his aunt and uncle had removed him from public school lest he "encounter someone who lives a lifestyle against the teachings of the Bible." I was brought in as an English, Spanish, Latin, and math teacher for Thomas, and the family adored me as a positive role model for him to aspire toward. *Ah, the irony*, I thought, chuckling. *Hiding in plain sight.*

Given his carefully curated and censored home life, Thomas didn't have any friends. I tried to structure my lessons around informality as best I could, giving him at least a taste of friendship. Standing out in middle school usually meant standing apart, and I knew what Thomas was going through. So when my phone rang for the fifth time with an unknown number that afternoon, I decided to have some fun with it. If it made him laugh or elicited a rare smile, it was a small success.

After my two-hour session, I thanked his uncle, collected my $60, and sauntered outside to my parents' truck for the short ride home. As soon as I turned the keys, my phone rang again. It was the same number.

"What?" I grumbled loudly rolling my eyes and parking on the side of the sleepy wooded neighborhood. I was not in the mood.

"Hello, sir, my name is Hayleigh Shore," spoke the friendly voice. "I'm from the Harvard University Admissions Department. Is Matthew around? We'd like to schedule his interview."

The color drained from my face as I realized the number was not a telemarketer. Had I really just yelled at Harvard a half hour prior?

"One second, ma'am," I grunted. *Maybe she won't know it was me.* I placed the phone facedown on my seat-belted lap, took a moment to breathe, and then held the phone to my ear, smiled, and broke out my effusively friendly weatherman voice. I hoped I sounded different.

"Hello, this is Matthew!" I stated. She reiterated her introduction for me, noting that she had a tough time trying to contact me earlier on. *Deny, deny, deny*, I thought. It seemed to work.

I had only applied to three colleges. The first was Lyndon State, a tiny school in Vermont located near St. Johnsbury, which I had toured a month earlier. Situated on a hill overlooking alpine forests and the somnolent town of Lyndonville, it wasn't a name-brand school, but it had churned out some great broadcast meteorologists. Jim Cantore, the figurehead of the Weather Channel known for always being in the eye of the storm, had graduated from there in the 1980s. It was a perfect fit for me. I fell in love as soon as I set foot in their college newscast studio.

Despite my immediate draw to the campus and the culture, there were some red flags—the 99.4 percent acceptance rate was one of them, along with some of the statistics I was seeing. The cost of attendance—$26,000 a year—was another. I was on my way toward netting more than $90,000 in third-party scholarships, which would have mostly wiped out the cost of attendance over four years, except I knew Lyndon wouldn't give me much of a backup plan if broadcasting didn't work out. I was sure I would be happy there and I loved the students I had met, but I also had a future to think about.

Cornell, the second college on my list, seemed to be the best option. It had a world-renowned meteorology program and was an Ivy League school. Even after financial aid, however, I had calculated annual tuition to be around $37,000. A tremendous financial burden, even with the scholarships. That's when I spotted an article in the *Cornell Chronicle* from 2010: "Cornell has announced it will match the need-based financial aid for admitted students who are also accepted to other Ivy League schools."

That was music to my ears. I'd used the Walmart price-match guarantee before at supermarkets. Now I just needed to find an Ivy League school known for handing out piles of cash. How different could this be? Strategically, and, admittedly dubiously, I applied to Harvard. I evidently had glossed over the part that mentioned an interview.

—~—

The day after my unfortunate phone call, I found myself standing in the "Pit," a nickname for the lowered entryway in front of Sturgis's main office, waiting for Hayleigh. It turns out she wasn't just the scheduler. She was actually driving down from Cambridge to interview me.

"Excuse me, are you Hayleigh?" I asked a well-dressed woman walking by with a notepad. She seemed out of place.

"Matthew?" she asked. I smiled. So did she.

"The school's a bit of a maze, so I figured I'd meet you out here," I said. I could tell right off the bat I had made a good impression.

We strolled through the disheveled hallways past the teachers' lounge, where Mr. Mathews was snacking on a muffin at the copy machine. I had booked the adjacent conference room last minute that morning; after all, I didn't have much advanced notice. A box of holiday decorations and a leaking water dispenser stood in the corner of the room.

I waited for her to sit before plopping down on a chair myself, the two of us angled toward one another at one corner of the sprawling conference table.

"Thank you again for driving all the way down here," I said, adopting a refined but sincere weatherman charisma. "It's like pulling teeth to get my dad to go up to Boston just for an airport run." She laughed.

Our conversation flowed naturally and smoothly; I pretended I was speaking with one of my mother's friends. I was better with adults than I was with my peers anyway. Adults made sense to me. *Call of Duty* and memes did not.

"So what's your favorite book?" she asked. I replied without hesitation.

"*The Giver* by Lois Lowry." It was a dystopian novel that explored the perceived satisfaction a human would have in the absence of choices and consequences. I had found its symbolism deeply evocative. "Have you happened to read it?" I asked. She seemed taken aback.

"I'm actually friends with Lois," she said, her eyes lighting up the same way mine had when she'd asked me about waterspouts a few minutes prior. I had a good feeling about this.

Thanksgiving passed and a deep freeze settled over the winter landscape. The only trees that weren't naked were the scrub pines everywhere on Cape Cod. I continued my daily grind of school and working forty hours per week, including at an ice cream shop, as a paid weather columnist at the *Barnstable Patriot*, and tutoring and busing tables at a country club down the road. I was happiest when I was busy.

In January I picked up a new tutoring client from Chris, a friend of mine, who referred me since he wasn't well-versed in precalculus. It was a Plymouth South High School student who was failing senior math, and it was the last credit he needed to graduate. *Piece of cake*, I thought.

I scheduled my first session on a Tuesday evening after classes, printing out directions from MapQuest and driving three miles to the Ponds of Plymouth, the subdivision where their neighborhood was located. A light snow was falling, blanketing the ground and muffling sounds to leave a peaceful noiselessness. I double-checked the address, parked at the end of the driveway, and marched to the front step, clip-board in hand.

"SHUT UP!" I heard as screams echoed inside the house. It sounded like two men brawling. I hadn't even rung the doorbell yet. Dogs were barking frantically. I texted my parents the address, took a deep breath, and knocked. My tire tracks had disappeared into the snow.

A man in his early fifties opened the door. He was about six feet, four inches tall, looked to weigh a little over 220 pounds and, by the look of his facial expressions, he wasn't happy to see me.

"Okay, what's your hourly rate?" he asked, angling his body to block an antsy Rottweiler from escaping. No greeting, no pleasantries.

"Thirty," I said, trying to look stern and formidable in my fuzzy blue jacket. I definitely wanted to be taken seriously here. The man, who I imagined to be the student's father, wrote a check and handed it to me.

"This isn't going to work, so here's something for your time," he said, his eyes darting around wildly. I could tell he was flustered.

"Excuse me?" I replied, confused.

"He's barricaded himself in his room and refuses to do any work," the father said aggressively. "He's not coming out. I'm at my wits end. If he wants to drop out, he can. I'm done. Sorry to waste your time." The door slammed in my face.

I trudged back through the snow angry and $30 richer. *That family is making a huge mistake*, I thought. *Oh well. I tried. I am* not *getting eaten by a Rottweiler today.*

I climbed back into the truck and began driving home, the peaceful snowfall a stark contrast from the household crisis I had just parachuted into. As I drove farther from the house and turned onto Long Pond Road, a nagging guilt began gnawing at me. I thought back to all my favorite teachers—Mrs. Runyon, Mr. Phillips, Mrs. Cardin, Mr. Carspecken, Mrs. Yalden. What would they do? I called my aunt, a teacher, for advice.

My eyes narrowing, I whipped the truck around, driving back to house and parking at the top of the driveway this time. I banged on the door and rang the bell until the father once again answered. He was sweating and breathing heavily.

"I'm not leaving until you give me thirty minutes," I said. I didn't even know his son's name. "You let him drop out and you're going to regret this for the rest of your life. I'm not going to let that happen."

Before he could say no, I stepped inside and stood on the welcome mat, snow melting off my boots and dripping onto the floor.

"Wait here," the father said, clearly annoyed at my presence. He climbed a flight of stairs, leaving me to fend off Mr. Rottweiler and his unlikely friend, Mr. Pug. *They're more afraid of you than you are of them*, I repeated to myself.

"Ian, open up," his father yelled, pounding on a door. I couldn't make out a series of stifled shouts that followed. Seconds later, a full-blown screaming match was ongoing. I ascertained that the son, apparently Ian, had barricaded himself in his bedroom by placing a bookcase in front of the door. The hoarse, guttural roars became violent.

Suddenly, I heard the crash of splintering wood. The dogs, who had been harassing me seconds before, scurried out of sight. I looked anxiously up the stairs in horror, hearing Ian launching verbal grenades at his beleaguered father. Seconds later, the father reappeared at the top of the staircase.

"You'll have to go talk to him," he said, brushing past me. *Why am I doing this?* I thought.

I climbed the carpeted staircase, still in my blue hand-me-down jacket, and turned down the hallway. A bedroom door was hanging from its hinges, with books and shattered glass surrounding a toppled bookshelf. I stepped carefully over the wreckage to find an oversized, gangly teenager spawned across his bed, fuming. He turned to face me. It was Ian Wilson—one of my sixth grade bullies.

"Um . . ." I started, unsure how to begin. I had *not* been expecting this. "I remember you. You probably remember me. Forget that. I want to see you graduate. You owe it to yourself."

I was met with an angry scowl and a look of suspicion. For a moment, my sixth grade fear returned. I pressed onward.

"Give me thirty minutes, and if you feel it's not working, I'll leave. But give yourself a chance." He shrugged.

"Take a few minutes and then I'll see you downstairs," I said. I turned and walked out of the room without giving him time to respond.

Seven minutes later, I was staring absentmindedly at a blank piece of lined notebook paper on the wooden dining table in the Wilsons's kitchen. Silence engulfed the room, akin to the refreshing calm that sets in after a line of storms. To my surprise, I heard footsteps on the staircase. Ian had decided to take me up on my offer.

The first half hour passed quickly—so much so, in fact, that I neglected to tell Ian when time was up. He was too engrossed in converting between standard form and vertex form and trying to plot a parabola to notice. I quickly deduced that he knew more than he thought he did. What he lacked was confidence, however; building that back would be my first step.

We worked for two hours that night. My job was one of trying to bring equations to life—what did the lines, curves, and numbers actually mean? And how could they be visualized? At the end of our session, Ian quietly asked, "Are you free tomorrow?" It was a sentence that atoned the past; I could tell he was grateful.

As the weeks passed, Ian's aptitude and morale grew. He took pride in demonstrating what he learned, nabbing an 82 percent on his next quiz. B's replaced D's. Lightbulbs replaced furrowed brows. And—in the end—he graduated.

At the time, it seemed insignificant. Ian was a tutoring client, and I had done my job—teach math. I was seventeen years old. It just so happened that I briefly entered the equation at what, for him, proved a crucial inflection point in his life. A small nudge in the right direction would inevitably have an enormous bearing on his future.

It made me wonder how many of those cascading ripples I had surfed. Which chance encounters had shaped my life? Was I even aware of their impacts? Was it Mrs. Runyon, the eighth grade science teacher who was the one person that year who assured me it was okay to love meteorology? Was it Jan, the cashier at Shaw's Supermarket whose checkout line I had been passing through since preschool, and who, despite not always having much, gave everything she had to help everyone who needed it? Or was it Mrs. Monska and Mrs. Findley, the ladies in the front office at Indian Brook Elementary School, with whom I had shared jelly beans every morning?

Looking back, it was all of them and more. I firmly believe there's something to be learned from every person that enters one's life. I'm just fortunate that those who have been a part of mine have had a lot to teach.

My interactions with Ian and Thomas also illustrated the empowering role that education and confidence could have. It's a moral I've carried with me to adulthood. Nowadays, I embrace my role on all platforms as that of a teacher. If my viewers, followers, readers, and listeners learn one new thing or end their days feeling just a little bit smarter, then I've done my job.

—w—

The next month or two of senior year passed by uneventfully. I spent most of what little spare time I had working on scholarship applications, scouring the internet for textbooks, or daydreaming about the trip I hoped to someday take to Oklahoma. I got my hands on a copy of Thomas Grazulis's 1,400-page *Significant Tornadoes*, which chronicled every recorded F2 or greater tornado (on a 1–5 scale) between 1680 and 1991. (The F in F2 stood for the Fujita Scale, named after Tetsuya "Ted" Fujita, the late Japanese tornado researcher. F2 tornadoes contain winds of 111 mph or greater.) I knew I'd hear back from Harvard and Cornell on April 1, or Ivy Day, at 5:00 P.M. Eastern time. I wasn't nervous about getting into either school—I was more concerned about paying for it. When the day and hour finally came, I was busy watching WCVB NewsCenter5 at 5:00 P.M. in the basement.

"Aren't you going to check?" my mother badgered. She was anxious. I was devoid of emotion.

"Harvey's talking," I replied. I was watching the weather.

"Everyone is coming over," she urged. "Go check. *For me.*" She had invited my grandfather and my aunt Meg over for pizza. Cheers or tears, at least Papa might have some wisdom to offer.

Sighing, I climbed two flights of stairs and slipped silently into my bedroom, prying open the MacBook I had purchased four years prior after a summer landscaping for $6 an hour. I closed the door behind me.

"Congratulations," read the email from Cornell. *Check*, I thought. I planned to review their financial aid letter later. I wanted to get back to watching Harvey's 5:15 P.M. forecast!

I navigated to an email from Harvard containing my username and password for their admissions portal. *Copy. Paste. Copy. Paste. Enter.* A brochure-like display of diverse college-age students hanging over a text box greeted me. They were smiling. I was in.

Good, I thought. That was that. I had met my expectations. But I didn't feel like celebrating. It was only the beginning of what I knew was a long road to achieving my goals. And I wanted to get rid of my acne, too.

I checked the clock: 5:16 P.M. If I rushed back to the basement, I could still make it in time to see Harvey's seven-day forecast. I clicked my laptop shut and plodded down the stairs.

"Well?" my mother asked, beaming tensely as she stood in the kitchen with her famed "ove glove" oven mitt on. She was making cake. The glove had been a Christmas gift from my great-grandmother. Rumor had it she pulled it out of her cabinet after forgetting to go Christmas shopping.

"Rejected at Harvard, waitlisted at Cornell," I said matter-of-factly. My mother frowned, her expression softening as she motioned to comfort me. It was just a temporary lie. For now, I wanted to catch the end of Harvey's forecast. The less talking, the better. Besides—I'd tell them the truth later.

And I had a lot to think about.

Harvard

Against all odds, I wound up at my last choice school: Harvard. After financial aid, Lyndon would have been pricier, and driving out to Ithaca made me rethink Cornell (the suicide nets didn't really sell it either). In the end, I figured I could always transfer *from* Harvard, but I could never transfer *in*. Harvard was a golden ticket. I just had to write it myself.

Clicking the ACCEPT OFFER button was daunting. I was sealing my fate and matriculating into a university that didn't have a meteorology program. I didn't even tap the button myself. I tricked Noodles, my scruffy and short-tempered dog, to paw the trackpad for me.

Summer passed quickly and, before I knew it, move-in day had arrived. I was paired up with a friendly math whiz from Delaware and assigned a somewhat opulent apartment in Apley Court; it even had a fireplace (and ample space for my green screen). I met my advisor three days in and presented him my four-year color-coded course plan on poster board. It wasn't conventional, but I was sure I could wrangle together an atmospheric sciences education.

Doing so would require a special concentration, or petitioning Harvard to grant me my own major. A concentration in atmospheric sciences had never been done before. Special concentrations were awarded an average of once per year, and the vast majority of applications were rejected. Besides, I couldn't apply until the spring semester, and even then, it would be a long uphill battle.

Culture shock set in fast at Harvard. At Sturgis, I was the smart kid. Now I was average, if not a bit below. My lab partner in physics was a bona fide prince, and my roommate received a $500 allowance from his parents every two weeks. Students used words like *apropos* and *intersectional* as if their lives depended on it, and some went so far as to use *an* in front of *historic*. I couldn't tell who was putting up a facade or if anyone was being genuine. I was just little old me.

The proctor of our tiny twenty-six-person dormitory organized frequent outings and get-togethers to help the student bond. Our first began by introducing ourselves with our name, where we were from, and our PGPs.

"I'm sorry, but what's a PGP?" I asked. I had the misfortune of being selected to speak first.

"Preferred pronouns," the proctor said. "It's what you *identify* as."

Identify? I assumed it was one of those ice breakers like the ones at camp where we had to use the first letter of our first name. I thought for a minute, and then blurted out my answer enthusiastically.

"I'm Matthew. I'm from Cape Cod, and I'm a natural disaster because I chase storms."

Crickets. Two dozen students peered back at me blankly, some squinting as if I had said something offensive. *Great*, I thought. *I've already screwed up*.

Following the advice of my student advisor, I signed up for CS50, an introductory computer science course. He told me it would give me the coding prerequisites I needed for some of the physics classes I'd inevitably be taking.

"Take it pass-fail," my advisor said. "Trust me."

On the first day of CS50, I and nearly a thousand others crowded into Memorial Hall, an auditorium-size amphitheater tacked onto the end of Annenberg Cafeteria, a Hogwarts-style freshman dining hall built in the 1870s. The four-story theater was jam-packed, with boom cameras and recording equipment jutting out in between the hordes of students. The lights eventually dimmed and a DJ began pounding away at a soundboard.

"This. Is. C. S. FIFTY," a voice boomed as colorful lights danced. A balding man, presumably the professor, strutted out in tight black jeans and a dark V-neck T-shirt. Students clapped. A pair of assistants wheeled out a massive sheet cake for display.

We learned binary, the language of computers—zeroes and ones. Simple enough. I could manage that. The second class was equally straightforward, and the third was reasonable. Once the deadline to drop CS50 passed, the level of difficulty skyrocketed. I could hardly even begin each assignment, never mind get results. I flocked to office hours every night from 9:00 to midnight alongside two hundred other students, but, with limited staffing, I was lucky if I could get a single question answered.

My other classes were going better, but not by much, and I couldn't shake the feeling that Harvard had made a mistake. The other students weren't like me. I was gaining weight, my face was riddled with acne, and I hardly felt like leaving my dorm. I was at a school that didn't offer any meteorology courses. I knew that even if Harvard hadn't made a mistake, I had.

Things reached a low four weeks into the semester in late September, when I ventured to Jefferson Laboratories, a maze-like building where I'd have my first midterm exam. Little did I know the exam had been rescheduled—I had somehow missed that email.

Naturally, I never found the testing site: it didn't exist. Instead, I wandered the building, whose rooms seemed like they'd been numbered by a drunken cross-eyed lizard. I passed room 453 despite there being only three floors. Before long, I was hopelessly lost. And I was losing precious time locating the exam room.

Nine forty-five became 10:00, which became 10:10. Eventually, I gave up. That was it—the doors were doubtlessly closed and the exam had started. *Pack it up*, I thought. *You're done. Go home. This is it.* I was on the second floor of the sinuous building and no one was around. I slumped against the wall, tossed my backpack on the floor, and began to tear up. I was done. It was my first time in life resigning to the universe. I had fully given up. My heart sank.

A dusty beam of yellowed sunlight glinted off a brass nameplate affixed to the door of a faculty office opposite me. Something about it caught my eye. It read ERIC J. HELLER.

Instantly, my mind flashed back to the Wednesday newspaper half a decade earlier. Could this be the same kindhearted reader who complimented my article? I knocked on the door, and a friendly woman about thirty years old answered. After a short exchange, I learned that it was *the* Eric J. Heller. Apparently my random reader from 2013 was a world-renowned physicist at Harvard. With nothing to lose, I scheduled an appointment to meet him.

~

October rolled around and the leaves began to change. I was just barely treading water in CS50, but I had made a friend, Martin, with whom I spent every night working on problem sets. We were both struggling, but at least we were struggling together. Our study sessions usually concluded with a 2:00 A.M. game of Ping-Pong in Apley's basement. I began to feel less lonely.

One day around Halloween, Martin disappeared without a trace. None of the other students knew where he went. His dorm room was vacant, the hand-printed callout glued to his dormitory door ripped from its perch. Even my faculty advisor, who had also overseen Martin's academics, refused to divulge where he'd gone. It was rumored to be an academic integrity issue. I was sad not only because I had lost a friend but also because he had helped my adjustment to Harvard without even knowing it. Maybe Martin needed someone to do the same, even if only to listen.

When it came time to meet with Dr. Heller, I didn't know what to expect. I ironed my button-down shirt, as I was eager to make a good appearance. His assistant, Roel, greeted me at the door.

"Rick's running a bit behind," he said with a smile. "He should be ready soon."

About five minutes later, the door to Dr. Heller's office creaked open. An older man in his late sixties or early seventies shuffled out. His face was staunch, adorned with wire-rim glasses and a mustache. As soon as he saw me, he broke out into a smile.

"Matt!" he said eagerly. "Come on in!" It reminded me of the opening scene in *Willy Wonka and the Chocolate Factory*, in which a stumbling, haggard-looking Gene Wilder suddenly keels over, somersaults, and surprises the crowd with his verve. I followed Heller into his office.

Our discussion was far from conventional academia—we talked about life in Plymouth, Heller's extensive research in physics and chemistry, and his TED talk on rogue waves. He asked me about my passion for weather, which launched us into a lengthy chat about outflow boundary-related waterspouts.

"How do you plan to study meteorology here?" he asked.

"Well I hope to orchestrate a special concentration, but that's going to be a challenge," I said.

"What would that entail?" he asked.

"I'd need to have all four years' worth of classes planned out, a bunch of letters from people . . . I'd have to find an advisor," I explained. The process was lengthy and arduous.

"You need an advisor?" he asked, beaming. "I'm your guy." It took a minute for the gravitas of what he'd just said to sink in.

I'd found an advisor? I thought after a moment. *Just like that*? Finding an advisor—someone to accompany oneself on a four-year-long academic journey—was supposed to take months, if not a year or more to find. Had he just offered twenty minutes into our first meeting?

"I'm sorry?" I asked. I was still in disbelief. He was signing up for a years-long commitment.

"I'll do it!" he said. I'd heard correctly.

I was in shock. Everything had been going wrong, and yet suddenly the most important item had fallen into place. Things were suddenly looking up.

—~—

"I found an advisor!" I announced eagerly to my faculty advisor, who was in charge of overseeing my progress during freshman year. He seemed skeptical.

"Who?" he asked. I could tell he was cautiously optimistic, but excited for me. He was rooting for me.

"Eric Heller," I said, before delving into the story of how we had first "met" through my newspaper article. I was still surprised that someone of Heller's intellect and stature read the newspaper. I didn't see myself as worthy of his time. Who else might be quietly paying attention to me?

I spent the next several months working tirelessly on the forty-page special concentration application. That meant procuring an endless stream of statements, faculty letters of support, and devising a rigorous educational plan. I hunted for resources, discovering a little-known $4 million recording studio in the basement of Harvard's Widener Library, complete with a green screen. I befriended the studio manager, who offered me a weekly Friday slot I could use to practice my on-air delivery.

I opted to cross-enroll at MIT during the second semester of my freshman year. I was still lonely, but at least now I was busy with atmospheric dynamics. The course, taught by a grandmotherly Italian woman and a man who reminded me of Simon Cowell, proved my first taste of the "under the hood" equations governing the atmosphere. I was hooked.

—

I got the call three weeks into sophomore year, on September 14, 2016. It was Tessa Lowinske Desmond, a professor and member of the Committee on Ethnicity, Migration, Rights at Harvard. She also ran the special concentration program, which only consisted of four or five students. I didn't recognize the phone number, but this time, I picked up.

"I wanted to let you know that the committee on special concentrations met and has approved your proposal," she said, pausing for a moment. "You officially have a special concentration in atmospheric sciences."

I was dumbstruck. Harvard had been around for almost four hundred years; this hadn't been done before. Now they were entrusting me to do it. Even writing the proposal had been an uphill battle—the head tutor in the Earth and Planetary Sciences Department had refused to offer his support, telling me I was too young to know what I wanted to do. I wrote an effusive and vehement rebuttal to his response and, to my surprise, the committee had chosen my side. I was ready to hit the ground running.

I registered for a number of graduate classes in atmospheric dynamics, since Harvard didn't have any for undergraduates. My first was dynamic meteorology: an equation-based physics course that required fluency with multivariable calculus, differential equations, and abstract linear algebra. Usually it was the last course required of seniors before graduation. I had only taken up through Calculus II, but if it meant the opportunity to get my hands dirty in the field, I'd teach myself multivariable calculus on the go.

I doubled up on courses at MIT, petitioning Harvard to cram extra classes into my roster. Scheduling wasn't always cooperative, so if a course was offered, I did everything I could to get in on it. On Thursdays I had to make an appearance in three classes in an hour—one of which was at MIT, which was two miles away. I could usually be seen sprinting through Harvard Yard, splitting through crowds of tourists with a bagged lunch in hand en route to catch the M2 shuttle.

I hardly had time to sleep but, for the first time, I felt like I was on the right track.

Chasing my Namesake

It was a Tuesday night in early October 2016, which meant I had lab for Physics 12B. Much to my dismay, there were no beakers, test tubes, or stuff to blow up. Instead, we were working with Arduinos, tiny circuit board–like mini computers, to build a radio antenna. Our professor had hidden an electromagnet in the classroom, and it was our job to find it.

The challenge seemed interesting enough, but my attention was elsewhere—Hurricane Matthew was bearing down on Haiti with 145 mph winds. It had rapidly intensified over the weekend and become the first Category 5 Atlantic hurricane in nine years; the last Category 5 was Hurricane Felix, which lashed Nicaragua and Honduras in 2007.

I had forecasted hurricanes before, but this one felt different. There was a certain angst in the voices of news anchors reporting on it, and weather Twitter was abuzz as each new computer model run came in. They were painting an increasingly dire picture for Florida, where a state of emergency was in effect. Closures were announced at Disney World for only the fourth time in the park's forty-five-year history.

I thought back to my bucket list—"stand inside a storm with my own name" had yet to be scratched off. So did "visit Kericho, Kenya; chase waterspouts in Key West; document thundersnow; and make a best friend." I had to cross at least one thing off that year.

Hurricane names are predetermined, organized on a six-year rotating list. I figured I could just wait until the next reincarnation of Hurricane Matthew in 2022, but that would be iffy. Convention has it that a storm's name is retired if it causes damage or loss of life to the extent that future use of the name would be insensitive. Names like Katrina, Andrew, and Wilma were struck from the list, for instance. This might be my last chance to meet my cyclonic namesake.

Giving into a sudden urge, I decided to peruse flights—there was one on Delta leaving Boston at 10:35 A.M. Thursday that would get me to Daytona Beach, Florida, at 4:09 P.M. Hotels were cheap, too, only $100 for a Marriott on the barrier islands. I didn't have classes on Friday either, so if I went, I'd only be missing two lectures. *This could actually work*, I thought.

"Everyone understand the instructions?" the teaching assistant asked, snapping me back to attention. I realized I had been daydreaming about hurricanes and missed everything she said. All the other students were bobbing their heads up and down. I nodded, too.

Rations, water, goggles, batteries, pretzels, peppermint gum, camera gear, math textbook. It was Wednesday and I was running through my mental packing checklist. I had broken down and bought tickets hours after wrapping up my lab the previous night and, much to my mother's dismay, I was flying to the Space Coast of Florida to put myself in the heart of the storm.

While I had never chased a hurricane before, I knew what I was up against. If the hurricane's eyewall, or the innermost ring of thunderstorms within which the storm's full fury resided, was to move ashore, it would spell utter destruction. Neighborhoods would be wiped out, a massive storm surge would sweep inland, and the area would be uninhabitable for weeks or months. A Hurricane Katrina–like impact wasn't out of the question. I was jumping into the deep end of the pool.

My forecasts were a little less dire, but still severe. That offered some sense of security, but not much. If I wanted to stay safe, my game plan would be simple: run from the water, hide from the wind.

Steering currents seemed like they'd tug Hurricane Matthew on a more northerly track, paralleling the coast and scraping the shoreline. That meant gusts of 100 mph or more, but that the worst of the impacts may be relegated to Florida's near-shore waters. It was a nail-biter but, no matter what, I'd have a front-row seat.

Hurricanes are heat engines. They derive their fury from warm ocean waters in the tropics, where sea surface temperatures routinely hover in the mid- to upper eighties between July and October. Hurricanes and tropical storms fall under the umbrella of tropical cyclones. They can be catastrophic, but they have a purpose—some scholars estimate they're responsible for as much as 10 percent of the Earth's annual equator-to-pole heat transport.

Hurricanes are different from mid-latitude systems. So-called extra-tropical, or nontropical, storms depend upon variations in air temperature and density to form, and feed off of changing winds. Hurricanes require a calm environment with gentle upper-level winds and a nearly uniform temperature field. Ironic as it may sound, the planet's worst windstorms are born out of an abundance of tranquility.

The first ingredient is a tropical wave, or clump of thunderstorms. Early in hurricane season, tropical waves can spin up on the tail end of cold fronts surging off the East Coast. During the heart of hurricane season in August and September, they commonly materialize off the coast of Africa in the Atlantic's Main Development Region. By October and November, sneaky homegrown threats can surreptitiously gel in the Gulf of Mexico or Caribbean.

Every individual thunderstorm cell within a tropical wave has an updraft and a downdraft. The downward rush of cool air collapsing out of one cell can suffocate a neighboring cell, spelling its demise. In order for thunderstorms to coexist in close proximity, they must organize. The most efficient way of doing so is through orienting themselves around a common center, with individual cells' updrafts and downdrafts working in tandem.

When a center forms, a broken band of thunderstorms begins to materialize around it. Warm, moist air rises within those storms, most rapidly as one approaches the broader system's low-level center. That causes atmospheric pressure to drop, since air is being evacuated and mass removed. From there, the system begins to breathe.

Air moves from high pressure to low pressure. That vacuums air inward toward the center. Because of the Coriolis force, a product of the Earth's spin, parcels of air take a curved path into the fledgling cyclone's center. That's what causes the system to rotate.

Hurricanes spin counterclockwise in the Northern Hemisphere, and clockwise south of the equator. Though the hottest ocean waters in the world are found on the equator, a hurricane could never form there. That's because the Coriolis force is zero on the equator; there'd be nothing to get a storm to twist.

As pockets of air from outside the nascent tropical cyclone spiral into the vortex, they expand as barometric pressure decreases. That releases heat into the atmosphere, causing clouds and rain. Ordinarily that would result in a drop in temperature of an air parcel, but because it's in contact with toasty ocean waters, it maintains a constant temperature; it's heated at the same rate that it's losing temperature to its surroundings. As long as a storm is over the open water and sea surface temperatures are sufficiently mild, it can continue to extract oceanic heat content.

Rainfall rates within tropical cyclones can exceed four inches per hour thanks to high precipitation efficiency. Because the entire atmospheric column is saturated, there's little evaporation to eat away at a raindrop on the way down. As a result, inland freshwater flooding is the number one source of fatalities from tropical cyclones.

The strongest winds are found toward the middle of a tropical storm or hurricane in the eyewall. The greatest pressure gradient, or change of air pressure with distance, is located there. The sharper the gradient, the stronger the winds. That's because air is rushing down the gradient. Think about skiing—you'll ski faster if there's a steeper slope.

When maximum sustained winds surpass 39 mph, the system is designated a tropical storm. Only once winds cross 74 mph is it designated a hurricane. Major hurricanes have winds of 111 mph or greater and correspond to Category 3 strength. A Category 5 contains extreme winds topping 157 mph.

Since the winds are derived from air rushing in to fill a void, or deficit of air, the fiercest hurricanes are usually those with the lowest air pressures. The most punishing hurricanes and typhoons may have a minimum central barometric pressure about 90 percent of ambient air pressure outside the storm. That means 10 percent of the atmosphere's mass is missing.

Picture stirring your cup of coffee with a teaspoon. You know that dip in the middle of the whirlpool? The deeper the dip, or fluid deficit, the faster the fluid must be spinning. Hurricanes are the same. But what prevents that dip from filling in?

Hurricane eyewalls are in cyclostophic balance. That means a perfect stasis of forces makes it virtually impossible to "fill in" a storm in steady state. Because of their narrow radius of curvature, parcels of air swirling around the eye experience an incredible outward-directed centrifugal force that exactly equals the inward tug of the pressure gradient force. That leaves them to trace continuous circles.

If you've ever experienced a change in altitude, such as flying on an airplane, or even traveling to the top of a skyscraper, you probably noticed your ears popping. That's because they were adjusting to the drop in air pressure with height. Now imagine all the air below that height vanished. That's the equivalent air pressure in the eye a major hurricane. The disparity in air pressure is why a hurricane is, in the words of Buddy the Elf, "sucky. Very sucky."

Sometimes hurricanes undergo eyewall replacement cycles, which entail an eyewall shriveling and crumbling into the eye while a new eyewall forms around it and contracts, taking the place of its predecessor. This usually results in a dual wind maximum near the storm's center as well as a brief plateau in intensification.

In addition to the scouring winds found inside the eyewall, tornadoes, tornado-scale vortices, mini swirls, and other poorly understood small-scale wind phenomena can whip around the eye and result in strips of extreme damage. A mini swirl may be only a couple yards wide, but a 70 mph whirlwind moving in a background wind of 100 mph can result in a narrow path of 170 mph demolition. Their existence was first hypothesized following the passage of Category 5 Hurricane Andrew through south Florida in 1992, and modern-day efforts to study hurricane eyewalls using mobile Doppler radar units have shed light on their existence.

Within a hurricane's eye, air sinks and warms, drying out and creating a dearth of cloud cover. It's not uncommon to see clearing skies or even sunshine. The air is hot and still, an oasis of peace enveloped in a hoop of hell.

There's such a discontinuity between the raucous winds of the eyewall and deathly stillness of the eye that the atmosphere struggles to transition. The eyes of hurricanes are often filled with mesovortices, or smaller eddies a few miles across, that help flux and dissipate angular momentum into the eye. Sometimes four or five mesovortices can cram into the eye, contorting the eyewall into a clover-like shape. That makes for a period of extraordinary whiplash on the inner edge of the eyewall as alternating clefts of calamitous wind and calm punctuate the eye's arrival.

—

My flight to Atlanta was uneventful. I stocked up on fudge in Terminal D at Hartsfield-Jackson International Airport, skipped merrily to TGI Fridays, and flipped open my Calculus III textbook. Major hurricane or not, I had an exam the following week.

After a hearty lunch of chicken tenders and fries, I double-checked my hotel reservations. I had booked multiple just in case either property decided to evacuate. Daytona Beach was under a mandatory evacuation order, but both the Marriott and a neighboring Hampton Inn planned to remain open for guests to ride out the storm. The front desk at the Marriott told me they had a number of media personnel already make reservations.

Reality hit when I boarded my connecting flight to Daytona. I was one of five passengers on the plane. *Maybe there's a reason no one's flying to Daytona*, I thought. When I landed, the airport was packed with Floridians hightailing it out of dodge. After deplaning in Daytona, my aircraft headed back to Atlanta as a full flight.

I ordered an Uber and waited twenty minutes. Only three were still operating across the entire town. It was 4:00 P.M. and officials planned to shut all access to the barrier islands at sundown. My glasses fogged up in the sweltering Florida steaminess.

We passed through abandoned neighborhoods and boarded-up businesses, the driver bidding me farewell once we had made it deep enough into the ghost town. Sandbags blocked the front entrance to the Marriott, so I decided to walk around back. The parking lot was empty.

Angst started to set in when I couldn't find a trace of life at the hotel. After circling the building three times, I finally noticed a handwritten note taped to the front door. CLOSED DUE TO MANDATORY EVACUATION was scrawled across the crinkled page. The lobby was dark and lifeless.

I checked my phone. It was on 10 percent battery and the damp Florida heat was quickly draining what little charge remained. A small red *1* appeared next to the green phone icon. I had a voicemail.

> *This is Samantha from the Hampton Inn and Suites*, the voice said. *Due to extreme weather forecasted in the area, our hotel will be ceasing operations and shutting down. Your reservation has been cancelled and any charges will be refunded within five business days. Stay safe.*

A knot formed in the pit of my stomach. Both my plan and my backup plan had fallen through. My phone was now dead and I had no Plan C. *Maybe this wasn't such a good idea*, I thought.

I walked along South Atlantic Avenue, aware that I was only a few feet above sea level. I had $80 in twenties, but there was no payphone in sight. Besides, who would I call? There was no one around.

Thirty minutes elapsed before I saw someone in the distance. As the figure neared, I could tell it was an old woman with blonde hair and leathery skin. Surely she could help.

"Pardon me, ma'am, but do you know of any hotels that are still open?" I asked. I explained the situation. From her expression, I couldn't tell if I was about to get scolded for coming to Daytona Beach in the first place or welcomed into her living room.

"Sand and Surf," she said matter-of-factly, not batting an eye. She pointed south. "It's just over there."

I walked about eight hundred feet, happening upon the unimposing tan cement structure after several minutes. The shutters were painted blue to match the roof. *Not hurricane shutters*, I thought. The front door was locked, but I heard a voice in the back of the hotel. With my suitcase in tow, I decided to follow the sounds until I found a man sitting on a patio chair.

"Are you the manager?" I asked. He shook his head. He was wearing a stained white tank top and sweatpants.

"No. Manager left," he said, speaking in a heavy southern accent. "He said we could stay here if we looked after the place."

I recounted what I had gotten myself into, which prompted the man to launch into a story about growing up in Alabama and fending off tornadoes.

"You can stay with us!" he exclaimed. "We have an ocean view!" He showed me the room he and his wife had selected. It looked like it was about six feet above sea level. She was nowhere in sight.

"First floor on the ocean!" the man said, beaming proudly. He reached for a pipe and inhaled. Sea spray was already splashing the sliding glass doors. I eyed my smeared reflection; it glared back at me like a prescient specter. Something felt off. A half-drained bottle of rum rested next to the man's plastic chair. My gut told me to leave.

"Just be careful you don't *become* the ocean," I said, thanking the man before scurrying off. Despite my eagerness to witness a hurricane, I realized I was out of my depth. I had to get off the barrier island. I walked back to the street, anxiously checking over my shoulder periodically to make sure I was alone.

In the distance I spied a vehicle about half a mile away. This was my chance! I decided to sit on my suitcase, placing my arms on the back of my head and giving the best impression of Dorothy Lange's *Migrant Mother* I could muster. Maybe, just maybe, it would work.

As the van neared, I recognized a familiar red logo on it: CNN! The van slowed and eventually stopped in front of me. In the passenger seat sat a tan man wearing makeup and a crisp polo shirt. A coiled earpiece dangled by his left shoulder.

"You all right?" he asked. He explained that he and his photographer were retreating to Orlando before offering me a ride to the airport. I graciously accepted.

I walked around the back of the van to throw my suitcase in the cargo hold. Atop canvas and burlap bags and bundles of wire was a twenty-four-pack of Bud Lite.

—∿—

So this is what you get when you cross Walmart and a slumber party, I thought. Four draining hours had elapsed and, at long last, I was settled in for the night. CNN had dropped me off at the airport, where I was instructed to go to the police station. They sent me to David C. Hinson Sr. Middle School before I was told to take an Uber to Champion Elementary School on the other side of town. By the time I made it there, they had reached capacity, but the driver told me to get out anyway. He had to flee town, too. I handed him $60 and wished him the best of luck before weaseling my way into the storm shelter.

One hundred-fifty people were crammed into a 200 × 200 foot cafeteria room. The rest of the school was sealed off. There were people from all walks of life: a young family with kids from Houston who had just moved to town, an older retired couple afraid to lose power, a spattering of students from Embry-Riddle University. Mixed in among them, however, were a number of people who I could tell would be trouble.

The wind grew loud by midnight as lights switched off. I curled up in a ball atop a small patch of tiled floor in the corner of the room, using my camera bag as a pillow. I was worried my laptop and camcorders would be stolen while I slept. My Bose noise-cancelling headphones had already been swiped.

No sooner had I fallen asleep than I was awoken by a commotion. A wrinkled woman in flip-flops was shouting at a police officer standing near the door. She spat whenever she spoke.

"You have to let me!" she shouted in a frail but venemous, raspy voice that reminded me of sandpaper. Moments later it became clear what she wanted: a cigarette break. The officer calmly explained that the doors were locked and it was unsafe to venture outside, but she refused to back down. On the contrary, she slowly amassed a group of angry evacuees armed with lighters. I silently observed the altercation.

After a tense few minutes, the woman acquiesced before returning to her nest-like blanket on the floor. She reached into a backpack. Through the darkness I could just barely discern what she was searching for: a pill bottle. She shook four or five small white tablets into her trembling hand before swallowing them, slumping over, and falling asleep. I whispered a prayer for her.

My attention turned to a man sleeping about fifteen feet away from me. A small worm-like piece of string was draped next to him. With a start I realized the string was alive—it was a snake! It slithered beneath the man as if to snuggle up next to him. Moments later, he woke up shouting.

Things only got weirder from there. Around 4:00 A.M. I opened my eyes to find a woman absconding with my math textbook. My suitcase was unzipped. As she tiptoed away from me I cleared my throat. She turned around and caught my eye.

"I need it for my back," she whispered harshly, as if I had offended her, continuing to walk away before I could even open my mouth. She looked like Mama June from TLC's *Here Comes Honey Boo-Boo.* I wasn't going to mess with her.

—∾—

The wind progressed to a roar around sunrise as the inner rainbands of Hurricane Matthew moved in. I somehow still had a shred of cell service, so I decided to check the radar. The eye still had a few hours before it would make its closest pass. It looked to remain offshore, but it'd be a close call.

"OUCH!" an old man yelled, his voice laced with pain and irritation. I was leaning against a window near the southern end of the cafeteria, but craned my neck to see what was happening. I recognized the two parents from Houston I had met the evening before. They were apologizing profusely to the venerable man and his wife.

I later found out what had happened. The couple had a three-year-old daughter named Erin who, under ordinary circumstances, loved to "play sleepover." That entailed them dragging a mattress into her room at home and sleeping next to her. Apparently, the parents had told her that the hurricane shelter was really just an enormous game of sleepover they had orchestrated for her birthday. It seemed like a good way to avoid making her nervous about the storm. That backfired when Erin had taken it upon herself to initiate a 6:00 A.M. pillow fight; the old man with now-damaged eyeglasses was her first target.

By midmorning the winds had topped 70 mph as decorative trees in the school-adjacent courtyard crumpled into the breeze. Conditions inside were equally stormy; another attempted mutiny against the police officer ensued as a band of cigarette-wielding Volusia County residents demanded to smoke outside. They didn't seem to understand that wind-driven downpours would extinguish any trifling flame.

The ringleader of the attempted coup d'etat, the old woman from the previous day, snuck into the women's bathroom shortly after the verbal brawl. About twenty minutes later the floor to the cafeteria was sopping wet. The woman had intentionally clogged all the plumbing inside the restroom as a form of retaliation against the police and shelter personnel.

Later that day officials spotted her wandering around outside in the parking lot. No one knew how she had escaped. Somehow she seemed immune to the flying debris.

The day passed with more lunacy; one man pulled out a bag of what resembled marijuana and began diving it into little piles, which he sprinkled onto pieces of brown paper. Another woman, who appeared to be under the influence of some substance, spent her time doing lunges and yoga poses, sans pants.

I befriended the Embry-Riddle students during the afternoon, finding that most were my age and learning to be pilots. We chatted for hours about meteorology, aviation, and the intricacies of air traffic control.

Still missing my math book, that night I decided to protect my $6,000 in camera equipment at all costs. I slipped away from the group, ducked behind the floor to ceiling dividers behind the cafeteria's stage, scaled hulking piles of gym equipment, and swung, Tarzan-like, from metal pipes. Eventually, I walled myself into a soccer goal, hanging my sweat-soaked T-shirt from the netting to dry. The air conditioning had been on full blast for the prior thirty-six hours and I had been shivering all day. *Watch me be the one to get hypothermia in Florida*, I thought.

The third day began with blazing sunshine and a calm, still air mass bereft of any trace of Florida's trademark humidity. Flights had resumed operation out of Daytona International Airport but, with no Uber drivers back in town, I was forced to ride in the cargo hold of an Embry-Riddle transport van. The school had dispatched the vehicle to pick up the dozen students who joined me in the shelter.

While I was relieved to depart the confines of the elementary school cafeteria, I knew I would miss some of the genuinely good people I had met. As I exited Champion, I complimented Erin's mother on her dress.

I was amazed that, after days without a shower or a mattress, she could look so put-together.

"Hurricane or not, I make sure to bring me my fabulous," she laughed.

I was humbled by the kindness of the faculty and staff at Champion. While the police were paid for their time, Champion's principal was volunteering her own. She welcomed more than a hundred total strangers into her school and provided them with warm home-cooked meals and shelter. While I was disheartened about how most of the evacuees treater her and the school, I tried to convey my personal gratitude.

In the end, Hurricane Matthew remained just a few miles off the coast. Florida was spared the calamity it could have otherwise faced. The name Matthew would eventually be retired on the grounds of what transpired with its passage in Haiti, which resulted in more than five hundred fatalities and $3 billion in losses. The continental United States was very, very lucky, but not everyone was so fortunate.

While that means I'll never stand in the eye of a storm bearing my name, the trip proved valuable in other senses. It offered ineffable takeaways that guide how I approach my work today.

Nowadays, I work to inform and serve members of the public who may be evacuating from a storm. For three days in 2016, I was a hurricane evacuee. I witnessed firsthand why many families with children choose not to evacuate, and came to appreciate how many people simply can't, especially those who don't have the financial means or who face medical challenges. The discussion surrounding hurricane evacuations is an incredibly nuanced one, whether due to limited lead time, forecast uncertainty/complications, demographic and socioeconomic challenges, or physical logistics like gas shortages or the implementation of roadway contraflow. My time in Volusia County would make me a better scientist by making me a better human being.

We're Going to Need a New Windshield

Before I knew it, spring of sophomore year arrived at Harvard. The dingy gray of Boston's characteristic winter overcast gave way to a promise of clearer skies, and the birds and insects seemed to be taking note. Harvard Yard was a muddy mess as piles of melting snow left the ground sodden and waterlogged. The courtyard in my new upperclassmen dormitory, Currier House, was abuzz with activity as students flocked outdoors to toss Frisbees and play Quidditch. I had never been aware it was a real thing.

I was neck deep in the hustle of six classes split between Harvard and MIT and felt like I needed the time turner Hermione had in *Harry Potter and the Prisoner of Azkaban*. Four of the courses were mixed undergraduate/graduate classes and definitely presented a learning curve. By some miracle it was all working, and I wasn't about to question how.

Back in January I had received a campus-wide email advertising a free trip to Israel. My mother had conditioned me to perk up any time I heard or saw the word *free*, so naturally I paid attention. By March I found myself in Israel's Negev Desert, floating in the Dead Sea. It was then that I realized I could travel on Harvard's dime. I vowed to study abroad my senior year.

In the nearer term, my sights were set on storm chasing in May. I'd always dreamed of getting out in the field and doing it, but until then had never randomly found myself "in the neighborhood" of Oklahoma. Now there wasn't much holding me back—I had time to spare and possessed a valid driver's license—but I knew the price tag would be a hurdle. Fortunately, I had a few tricks up my sleeve.

In my original special concentration proposal, I had included "storm chasing" as a footnote under a four-credit mesoscale meteorology tutorial, which meant it technically counted for a sliver of a single credit. That in turn permitted me to call in one of the outside scholarships I'd won from Coca-Cola to fund it since it was, on paper, an accredited activity.

It was a stretch, but then again just about everything I was doing at Harvard was outside the realm of typical.

There was just one issue. The Coca-Cola award totaled $5,000 annually. While they'd be fine with me applying it toward storm chasing, the company had a policy that dictated checks be made out to an institution rather than an individual. That meant Harvard would get the money, but I had to convince them to turn those funds over to me. Something like that had never done before, but I figured it was worth a try. I sent an email to the Griffin Office of Financial Aid, hoping to inquire about bending the rules.

To my surprise, Harvard seemed more than enthusiastic about my endeavor. They knew field experience would be just as valuable as classroom instruction. I was put into contact with Amy Staffier, an associate director of financial aid. We met in person and discussed my proposed plan. It didn't take long before she cut a check.

With the funds in hand, I quickly set about plotting my journey west. My parents were adamant that I have a copilot, but being a department of one didn't afford many options. I wound up posting an ad on the Harvard Class of 2019 Facebook page offering a free trip to anyone open to an adventure. One person, a curly haired math major from central Massachusetts, replied.

His name was Aaron. He was a statistics major, lived in Cabot House, just across the Quad Yard from where I was, and was also from Massachusetts. After a few brief emails, we met for lunch. He seemed bubbly, good-natured, and easygoing. I explained the trip as best as I could.

"It'll be tons of driving, tons of waiting, lots of nothing, and a few really intense moments," I explained. He nodded. "We're talking eight-plus hours a day in the car." I had to make sure he knew what he was getting himself into.

"I've never been out there," he said, smiling. "It'll be something new."

I pivoted my laptop toward him, conjuring up images of radar hook echoes, schematic cross-sections of tornado-producing thunderstorms and maps of Texas, Oklahoma, and Kansas. He nodded silently as I explained the significance of each, my hands gesturing as if I was a marionette being controlled by an over-caffeinated puppeteer. He committed to coming for a week before having to fly to Los Angeles for a summer internship.

"Wide open spaces," the Dixie Chicks sang over the speakers of the 2014 Honda Ridgeline my parents had lent me. My green 2007 iPod Shuffle rested in the cupholder, connected to the speakers by a knotted aux cord. Only forty-two of my favorite songs had made it onto my storm-chase playlist, which meant the tracks looped every three hours. I didn't mind, even if I heard the same songs nine times each during the three-day drive from Boston to Oklahoma City. "Room to make big mistakes . . ."

It was May 6, 2017, and it was go time. May is the busiest month of the year for tornadoes, averaging 276 twisters nationwide during the month. The vast majority occur over the Great Plains and central United States as the dying throes of winter and blooming warmth of summer wage war amid a predictable seasonal clash. This year I'd have a front-row seat for all the action.

I had never seen so much nothing as I did driving on the Kansas Turnpike for the first time. Sure, I'd driven along stretches of interstate in the Northeast and New England without strip malls, fast food joints, or supermarkets, but this was *really* nothing.

There were no buildings, no hills, and no rivers, and I could count on one hand the number of trees I saw. The silhouette of a tractor stood against a backdrop of clement sunshine five miles in the distance. My ever-churning mind began to slow as a blissful peace descended on me. After a decade and a half fantasizing about someday making it out to tornado country, I had arrived. I felt like Dorothy and Toto would be waiting on the berm of the road to greet me any minute. *Home at last*, I thought.

—∾—

The month started slow without much in the way of tornado activity. A brief flurry of storminess occurred as the atmosphere perked up around May 11 and 12, but most of the storms were hailers. That was around the time Aaron arrived in Oklahoma City. Without much to do, we passed the hours driving around the area.

We stayed in Moore, a suburb of Oklahoma City on the south side of town that had been largely razed on May 3, 1999, when an F5 tornado tore through the city. Thirty-six people died as the mile-wide cylindrical buzz saw of 300 mph winds obliterated neighborhoods and stomped city blocks into rubble. It was the first time the National Weather Service had ever declared a tornado emergency, meteorologists wracking their brains on how to deliver enhanced wording that would convey the life-or-death nature of the situation.

Moore's terse history with twisters only continued from there. An F3 tornado swung through the city again on May 8, 2003, knocking down some of the homes that had just been rebuilt. Yet another high-end tornado—and the last EF5 to touch down nationwide in a decade—claimed twenty-four lives in Moore on May 20, 2013, prompting the issuance of another chilling tornado emergency reminiscent of a fateful afternoon fourteen years prior.

As soon as I rolled into town, it became clear that everyone—Dalia, the front desk associate at the La Quinta on Southwest 119th Street; Mark, the waiter at the Waffle House; Amy at the barbershop on North Broadway Street—had their own tornado story. I felt like I was standing on a solemn battleground. Residents knew to hold their breath during the month of May.

At one point, Aaron suggested we kill a few hours and watch a movie somewhere. He routed us a mile up the road to the Warren Theater. It had been used as a medical triage facility during the 2013 tornado. The scars of storms past were replaced with new construction and unkempt vegetation, but they were never really gone.

On May 15, Aaron and I chased hailstorms in the Texas Panhandle. I had only seen quarter-sized hail once before; now I was frolicking about as golf-ball-sized hail showered down from the sky onto a dirt road intersection in vast empty fields. Naturally, that resulted in a few triumphant bruises, but at least I was wearing a helmet.

The next day began in Guymon, a small farm town in the Oklahoma Panhandle. I knew it would be the first "real" chase day we'd have (and, as I'd later find out, the only good one in a near record dull season). Aaron and I munched on stale blueberry muffins from the hotel's continental breakfast while I pored over early morning data. *Today's going to feature some scratches and dents*, I thought.

By lunchtime Aaron and I were on the western border of Oklahoma, where a foreboding red bullseye had been drawn in midday severe weather outlooks. A rare PDS, or particularly dangerous situation, tornado watch was in effect. The sun was out, but things were about to get ugly.

Storms erupted around 2:00 P.M., a trio of small but puissant cells detonating like mushroom clouds in the volatile atmospheric setup. They were moving southwest to northeast at 30 mph. A half hour into their reign, tornado warnings began popping up. Fastidiously focused, I drove to the unincorporated town of Alanreed, made up of three half-deserted streets and a cemetery, and waited for the storm to our southwest to barrel through.

Sunshine gave way to light rain to heavy downpours, eventually with marbles of hail mixed in. Doppler radar indicated the rotation was about to pass us by, but when it did—nothing. No tornado yet. I decided to reposition ahead of the storm once again for another intercept.

That's when I realized my rookie mistake: I had actually fallen behind the storm. Once a storm is past, it's virtually impossible to sneak ahead of it once again. The next hour proved a futile fumble of traffic jams and receding clouds.

At 5:00 P.M., I accepted my blunder, resigning to my incompetence. The storm that had slipped by in Alanreed went on to produce a tornado just twelve miles away, which I had missed. But something in my gut told me the day wasn't over yet. It turned out I was right.

A new storm was forming in southwest Oklahoma, and, if we left immediately, we'd catch it. We raced east on Interstate 40, arriving in Sayre, a rural community just west of Elk City in western Oklahoma, an hour later.

We exited the highway just after 6:00 P.M., turning south and rolling along the undulating hills of Oklahoma State Highway 283. Heavy rain was coming down, but radar data implied it was about to end, and I was moments away from encountering the strongest thunderstorm of my life.

"Dear lord," I said to Aaron suddenly, who appeared equally awe-struck. As if on cue, the curtains of rain lifted, revealing a hulking spiral tower of black clouds slowly revolving to our southwest. A bell-shaped depression dipped out of its base, obscured by a film of rain and hail draped around it. Two arm-like appendages—channels of inflow cork-screwing into the fifty-thousand-foot-tall storm—were wrapping around each other like a curved staircase of brume to the stratosphere. We were watching a supercell. It looked villainous.

Within an hour, the cell would put down a deadly EF2 tornado just to our east in Elk City. The vortex was rain-wrapped, but that didn't stop me from angling closer in an attempt to sneak a peek. That meant driving into the heart of the storm.

"Aaron, now's a good time for you to reach under your seat and grab the safety gear I have for you," I said at 7:00 P.M. as we approached the tempest's core. I was ready, but Aaron thought I was joking.

"No, actually," I said after a moment, my voice more firm. The angry hiss of the elements seemed to want to force entry into the truck. "We're about to get some hail."

I had placed a hard hat, safety glasses, and work gloves beneath the passenger seat for an occasion like this; upon finding the items, Aaron seemed startled.

"How big is this hail going to be?" he asked, suddenly perturbed. "Like nickel size? Quarters?"

"Nope," I said, grinning demonically. "Softballs. We're about to lose our windshield."

I was right. And it was the best day of my life.

Anatomy of a Supercell

Supercells are the king of thunderstorms. They can tower ten or more miles high, spin like a top, and produce some of the fiercest weather on Earth. They are elegant and destructive, and beautifully terrifying. And they are a force to be reckoned with.

Supercells aren't like ordinary thunderstorms, which clump together into clusters or gusty squall lines. Supercells are small and potent. A single storm may only be five or ten miles wide, but in that tiny space could be packed destructive straight-line winds, softball-sized hail, flooding rains, and tornadoes.

What makes a supercell special is its isolation. While other thunderstorms may compete with neighboring storm cells for resources, supercells exist on their own. That gives them full reign of the surrounding environment—an untapped stockpile of fuel.

All thunderstorms have two main regions: the updraft, where warm, moist air rushes inward and upward, and the downdraft, within which rain (and sometimes hail) crash to the surface and drag down cool air from aloft. Ordinary thunderstorms develop vertically, meaning the downdraft eventually chokes off the updraft, spelling an end to the storm. So-called pulse thundershowers are common in the summertime, and may only last an hour or two.

Supercells are different. That's because they are sheared, or subjected to a change in wind speed and/or direction with height. That causes them to spin and lean, separating the updraft from the downdraft, a segregation that fosters longevity. Supercells can travel hundreds of miles, hopping state borders and spinning along for hours. During the infamous April 27, 2011, tornado outbreak in the southeastern United States, one supercell spun for seven hours, traveling 350 miles from eastern Mississippi to western North Carolina.

The supercell's rotating updraft is known as a mesocyclone, an area of spin a few miles across. It can resemble a barber pole, tilted southwest to northeast. Warm, moist air from the south spirals inward, tracing a helix as it climbs high into the sky. Air can race upward at speeds topping 100 mph, suspending hailstones as they grow to mammoth sizes. The strongest supercells can produce hail up to six inches across.

There are three species of supercell thunderstorms, and each offers different delights and hurdles for storm chasers. Classic supercells are the easiest to chase, since they frequently breed tornadoes and adhere to a textbook separation of their updraft and downdrafts. That makes it possible to more quickly locate where a tornado would form.

High precipitation (HP) supercells are enshrouded in rain. That's because the moisture-loaded storms feature curtains of precipitation that cloak the updraft. Chasing tornadoes in HP environments can be a dangerous game, since they're often not visible until the last minute.

Low precipitation (LP) supercells are arguably the most photogenic. They exist in drier environments with nothing to obscure the cylindrical structure of the updraft. LPs are sometimes tough to spot on radar and resultantly can be easily underestimated, but can drop baseball-sized hail without yielding more than a drop or two of rain. LP supercells, while visually stunning, rarely produce tornadoes.

At the base of a supercell's corkscrew updraft lurks the most ominous part of the storm. That's where a rotating wall cloud, or an eerie low-hanging, rugged cloud appears and sucks air into the storm. It's from that slowly orbiting mass of smoke-like clouds that tendrils of more concentrated rotation reach to the ground: funnel clouds. Once a funnel snakes to the surface, it's a tornado.

Some tornadoes are slender, photogenic ropes that dance playfully over open fields. Others are encased in rain, spanning two and a half miles wide and swallowing entire towns. As the rotating updraft churns round and round, it entrains a strip of rain, hail, and chilly air and tugs it around to the backside of the storm. This is the rear flank downdraft. That surge of cold, dense air in the storm's wake can tighten the updraft's rotation, squeezing and stretching it into a tornado. Sometimes the rear flank downdraft wraps all the way around a tornado, dragging curtains of precipitation around it. These hidden tornadoes are the most dangerous.

If a supercell is bearing down on you, conditions will change on a dime. You may first notice a distant, puffy white cloud roiling with plume-like updrafts, gradually appearing darker as it approaches. Its overhead anvil will extinguish the sun as the winds go still, faint grumblings of thunder interrupted by periodic chirps of anxious birds.

Eventually, the storm draws near, a light pitter-patter of rain quickly becoming a torrential downpour. The rain becomes colder, with fewer but larger rain drops splattering against the pavement. You might notice a raindrop bounce. *Bounce?* You ask yourself. Seconds later, the rain is replaced by a shower of pea-sized hail, with larger stones becoming mixed in. The sky takes on a phosphorescent green color. We're still not entirely sure what makes the sky turn green, but it's probably a product of evening orange light skewed blue through thick columns of precipitation in the tallest clouds. The different colors of light average to green.

The hailstones grow larger, bouncing higher and bombarding the landscape. Marbles become golf balls and tennis balls, hitting the ground and shattering in an explosion of icy shards. Eventually, the darkened skies lift, but meteors of ice still hurtle toward the ground. Some leave craters. They're now the size of softballs. Only a few are falling, but they're enormous.

Suddenly, you catch sight of a scuddy cloud about a mile away from you. It seems too low. Something's not right. All the other clouds seem to be rushing toward it.

Moments later, a pointed cone emerges from the smoky whirl of clouds, aiming at the ground. Below it, a tuft of dust is kicked up. After a few seconds, a full funnel has formed. It's a tornado—and it's growing.

Roadside Attractions

If you're a storm chaser, odds are you're accustomed to putting some serious mileage on your car. I quickly learned that a single week in May might feature visits to Nebraska, Kansas, Oklahoma, and Texas. Storm systems move, and that means I do, too. Most days feature three hundred miles or more of driving.

I've crisscrossed just about every paved highway in the Great Plains. I know where every speed trap is (well, *almost* every speed trap), where to find the most scenic picnic spots, and which gas stations have microwave ovens. I always know how many miles it is to the next town with a Chili's, and I'm never more than an hour away from one of my dozen or so secret stargazing spots.

By day, eight-hour drives can become long and monotonous, especially when I'm alone. If "America the Beautiful" was written proportionally, the amber waves of grain would take up three verses. They're endless. Playing a one-man game of "I Spy" isn't overly entertaining when all there is to see is crops and silos, and the forty-six-song playlist on my green iPod shuffle loops every three hours; streaming signals don't work very well out on the prairie.

My favorite road trip pastime, however, stems from the roadside billboards. Most advertisements on the Great Plains aren't for Coca-Cola or McDonald's. They're for casinos, gun shows, XXX video stores, bail bondsmen, farm equipment, the highest payout bingo game, and deep-fried pies. Those always spark a chuckle, but some billboards have a threatening aura.

WILL YOU BURN!? read one I passed in 2018, fiery-red text superimposed atop sinister crimson flames. HELL IS FOREVER screamed another. Others are even more ominous: GOD IS WATCHING. IS HE PLEASED? The Bible Belt is full of scary warnings along the side of the highway.

Some are bluntly specific, listing all the things they caution will get you a free pass to hell. Drinking in excess? Yep. Gluttony? Mmm hmm. I always roll my eyes when I see that one, as I'm usually busy plotting my next meal. Sexual immorality? The billboards warn against that, too. Frankly, hell doesn't sound like that bad of a time. Besides, you have to climb stairs to get to heaven, but there's a highway to hell. I wonder if that highway has billboards, too.

Once in a while, I pass a special billboard. SEARCHING FOR GOD? CALL NOW! it reads, a boldface 1-800 number emblazoned against a blue sky stock photo. Those are my favorite. I always call on the off chance that God picks up that day. I'm sorry to say, He hasn't yet.

There are some places too remote even for billboards, however. Those are my favorite. With nothing in sight save for distant rock formations or the blades of a far-off wind turbine, there's little standing in the way of a sunset as vibrant as it is pure.

The landscape of the Great Plains, while at first glance mundane, teems with beauty. Every rolling crest and dip of the prairie is part of a soothing rhythm, the drive akin to the gentle swaying of a ship at sea. The only difference? I'm not steering my ship into safe harbor like most captains. I'm out in pursuit of the storm.

Junior Year and Thundersnow

By junior year, I was halfway there and was most certainly living on a prayer. I had a deficit of sleep and a surplus of food. Free time was hard to come by.

Lacking the confidence, time, and, frankly, the know-how necessary to develop a full social circle, I gorged myself on work. If I was occupied with professional activities and earning money, I could easily defer my personal life. Besides, I didn't particularly want to confront that part of me. Sometimes I still don't.

Fortunately, it wasn't tough to stack my schedule. Academically, I was nearly doubled-up on courses, and it seemed like opportunities for side hustles were cropping up like dandelions. Leading up to the start of the year, a gentleman named Jason Samenow, the weather editor at the *Washington Post* and mastermind behind the *Capital Weather Gang*, emailed me out of the blue. "Your work is very impressive and I love your passion," he wrote. "Let me know if you might have any interest in working with us here in some capacity down the line." His brand had a cult-like following and was adored by fans across the Washington, DC, metro area and beyond.

We set up a call for the following week, and it didn't take long to figure out that Jason would be a superb mentor. He had grown what was once a casual weather blog and his side hobby into a popular brand that netted more than one hundred million reads annually. He invited me to submit pitches and try my hand at writing an article. My first piece was on lightning.

The year 2017 proved to be a banner one for weather coverage. Within two weeks of joining the "gang," I was writing every day, first about the Great American Eclipse of August 21, for which I flew to Nebraska, and then covering Hurricanes Harvey, Irma, and Maria, a trio of Category 4 lusus naturae that made landfall in US states and territories. It seemed like a relentless rampage of storms had followed me to my new gig.

Around the same time I began delivering tours of Harvard, working for a third-party company that orchestrated seventy-minute circuits for tourists eager to explore the esteemed university. I also did some tutoring at Boston University. The work was easy, and it was extremely profitable; each article netted $80–$120, I made an average of $60–$120 in tips per tour, and tutoring earned me another $50 an hour. I felt like I was getting a head start on my future.

I was also firmly into the more niche classes I'd be taking in atmospheric sciences, affording me plenty of time to dive into the nitty gritty. That didn't mean it was easy—one of my homework questions asked me to find air temperature and pressure as a function of radius and density in an imaginary drum hurtling through outer space and spinning at a given angular velocity. I didn't always get all the answers right (in fact, I never did), but I learned a lot through trying.

Dr. Heller was offering a general education class that semester, too, called Why You Hear What You Hear. Dr. Heller was the kind of person who could pick up a hobby and shortly thereafter publish groundbreaking research at the forefront of whatever it was he had tried. During my time at Harvard, that turned out to be acoustics. Within a year or two of kindling a curiosity, he had a thousand-page textbook out and was delivering classes on everything he had discovered. I hope to someday be like him.

My favorite class that semester, however, wasn't exactly a class at all. It was a one-on-one tutorial that had been approved by the Committee on Special Concentrations. The teacher was Daniel Davis, the laboratory demo guy from Physics 12B in charge of dreaming up visuals and experiments at the request of the professor. It turned out he had a PhD in atmospheric electrodynamics (i.e., lightning).

Davis had worked at the Boston Museum of Science as the resident expert on lightning and electricity, delivering shows to audiences and bravely standing inside the exhibit's Faraday Cage as it was zapped by a 20,000 volt arc of electricity. Before that, he had fired rockets into thunderheads at Los Alamos National Laboratory in New Mexico in hopes of triggering lightning strikes.

The tutorial took the form of weekly meetings on Fridays at 1:00 P.M.; Davis sent me an assortment of readings every Monday via Dropbox, which we discussed in depth during our sessions.

While they were all fascinating, one particularly captured my attention—a 1994 paper by N. Kitagawa and K. Michimoto, who examined a climatology of wintertime lightning events in Japan. They found that mature winter thunderclouds exhibited a tripolar electric field, with a region of negative charge at the mid-levels sandwiched between a broad positive charge at the top and bottom of the cloud. That got me thinking.

Every meteorologist and weather weenie yearns to experience thundersnow; it's one of the hard to check boxes that remains empty on most people's weather bingo cards and bucket lists. Thundersnow is exactly what it sounds like—lightning and thunder that accompany a heavy snowstorm. Since snow acts as an acoustic suppressor, only those directly beneath a strike hear its rumble.

Sometimes thundersnow events form just like ordinary thunderstorms, but with temperatures cold enough to support snow. That's inherently rare, however, since it's hard to get an unstable atmosphere and rising air when the mercury dips below freezing.

Thundersnow can happen in other ways, too. Elevated thunderstorms form when the surface is cold, but air a few thousand feet above the ground is mild. In other words, the thunderstorm is rooted in warm air half a mile or more above the ground. Any precipitation that falls to the ground can freeze in the shallow chilly air mass, but it usually accretes as freezing rain rather than snow.

Growing up, I noticed that thundersnow sometimes formed along eastern Massachusetts coastal fronts, or the dreaded rain/snow line. That's where a mild marine air mass brings rain at the shoreline while feet of snow fall just a few miles inland. In addition to being the bane of forecasters' existences, the temperature contrast sometimes enhances omega, or upward motion, enough to induce thundersnow.

The most meteorologically dynamic setup for thundersnow requires slantwise convection, the product of conditional symmetric instability. It sounds like a mouthful, but it's actually quite simple. Pockets of air aren't vertically buoyant, so they don't rise as they would in a normal thunderstorm. Instead, surfaces of constant potential temperature, a quantity that describes the temperature a parcel of air would be at if it descended to the surface, slope more steeply than angular momentum surfaces, derived from the rate at which parcels of air siphon into a spinning low pressure system. Tubes of air slosh diagonally up those density gradients and form bands of heavy snow and thunder.

During my tutorial, I also read about the Chicago thundersnow blizzard of 2010. That's when Jim Cantore, lead meteorologist at the Weather Channel, popularized the term with his viral in-the-storm freak-out. Apparently, the vast majority of cloud-to-ground lightning strikes during that storm were to tall man-made objects; virtually nothing on the surface was struck.

Davis taught me that, when a charge gathers, it ionizes the air around and induces an opposite shield charge that prevents an object from being struck. That wards off a lightning strike. But experimentally, he had found that, if air is blowing around an object at sufficient speed, the shield charge can be carried away, exposing that object to experience a discharge.

I began to pay attention to thundersnow events around the country, mapping the location of most strikes to either skyscrapers or wind turbines. I thought back to a pair of events I had experienced, including a blizzard on December 29, 2016, that I had chased to Oxford, Maine; I'd rented a room in an empty casino's hotel and dejectedly stared out the window all night long, only for lightning to strike the Prudential Center in downtown Boston.

A fierce meteorological bomb, or explosively intensifying storm system that lashed the New England coastline, brought a peppering of lightning strikes to southern New England on January 4, 2018. Most were to television and radio towers in Connecticut and west-central Massachusetts, though a few wind turbines west of Providence got zapped.

I discussed with Davis one particular winter storm that had hit a year prior, on February 8, 2017. It was the most prolific lightning-producing snowstorm I'd ever seen. Hundreds of thunderclaps came atop howling winds and rapidly accumulating snows, bringing five hours of lightning and thunder to my parents on Cape Cod. I had advised my father not to break out the snowblower during the storm due to the lightning threat. I even got lucky enough to enjoy one of the flashes from the roof of Harvard's Science Center; I was the only student with my own key.

Using what Davis had taught me, I began to hypothesize what was going on. Warm-season thunderclouds boasted strong electric fields that could spit out lightning seemingly incessantly. Broader, weaker updrafts in the wintertime, especially during slantwise convection in nor'easters, yielded more diffuse, meager electric fields.

Most of the time those fields weren't strong enough to generate lightning strikes of their own. But pointed manmade objects, usually taller than 800 feet high, could focus a snow cloud's lower positive charge enough to "go into corona" and spew a stream of electrons. That in turn would jump-start the growth of a lightning strike and heat the air ahead of it. The dielectric breakdown coefficient of air, or electric field threshold needed to get a spark to jump, decreases as air warms. That heating permits a strike's segmented growth to continue.

Bolts of thundersnow are typically positive lightning strikes with fewer return strokes, or pulsating flashes, and a lower peak current than warm-season lightning. They probably rectify charge imbalances between a cloud's lower positive charge and the mid-level negative region. Davis and I were especially interested in the role of wind in blowing away shield charges and leaving tall object susceptible to strikes. His previous laboratory work indicated the effect was strongest with winds over 35 mph. The theory also helps explain why wind turbines are more likely to get struck when their blades are spinning.

My time with Davis became fully dedicated to investigating the burning question that, until then, I hadn't been able to answer on my own. That's how the best science happens—through stubbornly, if not obsessively, pursuing the answer to a nagging question.

Someday maybe I'll get around to a more formal investigation into thundersnow or publishing a paper on it, but for now I can sleep just a little easier. Having sated my curiosity, there's one fewer question that'll be whispering through my consciousness at night.

Fire and Rain

So if anyone ever has any interest . . ." Bob concluded, looking over the sleepy classroom. The lights were off to allow easier viewing of his PowerPoint slide deck. The grad student to my right stifled a yawn. For the first time in Oceanography 166, I was rapt and attentive.

"I am!" I said, perhaps a bit louder than I intended to. My hand shot up instinctively even though there were only eleven students in the room. Bob, a visiting guest speaker from the Woods Hole Oceanographic Institution, had just invited us to volunteer on a research cruise to the Arctic the following summer. I was in as soon as he uttered the words *free trip*.

"Excellent," he said, smiling at me. "Shoot me an email and we'll talk." I was surprised, no one else was raising their hand. *Better for me*, I thought.

It was the end of junior year, and the finish line was finally in sight. I was weeks away from my May 2018 storm chase, had an internship lined up at the Weather Channel in Atlanta and, apparently, would be flying to the Arctic immediately thereafter. I'd also arranged to study abroad in the fall of my senior year. For the first time in a while, things were looking up . . . at least in the near term.

While I was beginning to see the light at the end of my four-year tunnel, burnout, social solitude, and imposter syndrome were taking their toll. It was also sensible to wonder if I was steering my career toward a dead end—the only job openings for an on-air meteorologist that I had seen on Indeed.com offered between $25,000 and $40,000 for full-time. It seemed the industry was hemorrhaging talent, rapidly losing viewers to the internet and facing a mental health crisis. An exodus of my colleagues into other fields had me wondering if I should be stitching together a backup plan. Would I be wasting a Harvard degree to get the kind of pay I'd net waiting tables part-time?

No matter where I looked, people were better than me at everything. Students in my graduate atmospheric dynamics class could churn through equations by the time I figured out which one I was supposed to use. Coding came effortlessly to them, yet I shuddered at the mere sight of the font Courier New. Some of my classmates were even getting married, and my closest companion was a cactus that I somehow managed to overwater (RIP Mr. Prickly). All the while, I questioned if I'd ever get to attempt dating down the road, and what my family, colleagues, and high school friends would think if they knew what I was hiding. *Career first for now, then personal life*, I thought. At least I had finally been prescribed Acutane and my face was clearing up.

I returned to the Great Plains for my annual engagement with the atmosphere upon the conclusion of classes in late April. I'd upgraded my equipment and was driving a brand new Honda Ridgeline I tenderly nicknamed "Ridgie." The first chase of the season involved an encounter with softball-sized hail near Susank, Kansas. Ridgie, now pockmarked and dented, had earned her stripes.

Bob and I reconnected over a Zoom call weeks later while I was in Woodward, Oklahoma. "Do you know what Zoom is?" he had asked via an earlier email, to which I responded that I did not (it was 2018). After downloading the software, I sat in my shag-carpeted hotel room at the Comfort Inn watching Bob give me a virtual tour of the US Coast Guard icebreaker *Healy*.

I didn't know what I was getting myself into, but I was fully on board. It sounded like the adventure of a lifetime. Before then, however, I had weeks of storm chasing ahead. And the atmosphere had plans to keep me occupied.

The atmosphere is the quintessential teacher. It always shows its cards, but never reveals what it's going to do with them. Getting the right answer requires astute observation, careful note-taking, close listening skills, and an open mind. Sometimes you learn more than you expect.

It's easy to chalk up the atmosphere to a series of rules and equations yanked straight from a textbook, but understanding the environment's caprice means getting to know its personality. Like any teacher, the lessons imparted aren't always conventional. Sometimes you learn about the subject material, and occasionally about life in general. Once in a while, you learn something you didn't know about yourself.

The month-long storm chase I had embarked on following my sophomore year brought to life the diagrams and charts I had rifled through for years. More importantly, the storms I hunted down and revered had a grounding effect. I later came to understand why—enveloped in an entanglement of chaos and fury, I briefly became the calmest thing in my environment. Ordinarily, I was my own self-contained natural disaster. Now, by comparison, I was a steady anchor point in the universe.

After that initial round of storminess on May 1, the atmosphere fell dormant. Storms avoided the Great Plains at all costs, any energetic upper-air systems diverting north and south, the flow bifurcating around a stubborn death ridge over the central United States. That dome of high pressure and sinking air squelched thunderstorm chances. I could hardly find any clouds.

Trevor, a friend of mine from weather conferences and a rising junior in meteorology at the College of Charleston in South Carolina, had elected to join me for a week of chasing a few months ago. He arrived smack dab in the middle of the dry spell. After five days of chasing sunsets and ice cream parlors, he grew bored and flew home early. I was ready to throw in the towel myself.

Then something unforeseen happened on May 11. I spent the afternoon sitting on Ridgie's tailgate, which I had parked next to a railroad crossing in Moore; it had become my default home base during chase season. My legs dangled as I scooped up a flimsy piece of Domino's pizza, ladled with chunked tomatoes and sliced jalapeños. Two freight trains had rolled by already. I counted 186 cars and five locomotives on the second one.

Absentmindedly, I flipped to Lightningmaps.org, which plots lightning strike that register on the National Lightning Detection Network's sensors. I was curious what the map looked like in Florida. Something seemed awry, though; there was a dense spattering of overlapping minus signs cumulating near Clarendon, Texas. The day was supposed to be sunny and dry. Maybe the data was outdated, but I had to double-check.

I flipped to radar—there wasn't a storm, but a slender yellow box around Donley County in the Texas Panhandle. Reflectivity mode on radar wasn't showing much, if any, rainfall, but echo tops indicated that a cloud had towered to 50,000 feet. The air was dry, with a hot breeze blowing in from the southwest. The sky was smoky.

I soon ascertained what was going on—a small grass fire had become a full-fledged inferno, releasing so much heat into the atmosphere that pockets of now-buoyant air were able to rapidly ascend. They had enough upward momentum that they'd broken the cap, flourishing in the frigid air 30,000 feet above the ground. Before long, the smoke plume had generated its own towering cumulus cloud, eventually culminating into a thunderstorm so tall it scraped the stratosphere. It was a pyrocumulonimbus, or fire-induced thunderstorm. I was flabbergasted.

Smoky thunderstorms were a thing of meteorological folklore. I'd heard of them occurring during the height of wildfire season in California before, but never during the spring in Texas and Oklahoma. Only a 10 percent chance of rain had been advertised that day, with a marginal risk of severe weather. Yet I found myself hastily shoving my pizza box shut, slamming Ridgie's tailgate into the locked position, and hopping the interstate bound for the western Oklahoma border. The atmosphere had a twinkle in its eye.

The soporific grumble of the pavement and intermittent trembles of Ridgie in the wind were mollifying as the sky grew pale. The formerly limpid sky looked like it had been coated in primer. Eventually, a tan complexion won out, ultimately becoming muddy and brown. It was like the world had lost its luster; the same thing had been happening to me at Harvard.

Veins of purple lightning leapt from the effervescent sky, which was textured with globules suspended from the storm's anvil. A variegated wall cloud was loitering in the distance, but I couldn't see evidence of rainfall anywhere. Still, I knew it was there.

By 5:00 P.M., I exited the highway and parked on a rural road near Elk City. *This place must be good luck*, I thought, thinking back to my first supercell a year prior. For hours, a combination of hybrid smoke and storm clouds wafted overhead, spitting out pinpoint lightning bolts after minutes of soft silence. Radar returns largely vanished by nightfall, but the electrified sky continued to sparkle. The heavens were palpitating.

At long last, and with an SD card teeming with hundreds of high-resolution photos that were, I daresay, museum quality, I began the long, nocturnal trek back to Oklahoma City. Darkness had fallen and stars were sparkling like shreds of tinsel.

An hour into my ride, I pulled to the side of Interstate 40 to, uh . . . "precipitate." There were no headlights anywhere to be seen; only the blinking red bulbs atop wind turbines gave any semblance of depth to the landscape.

Seconds into my performance, CRASH! From thin air, a webbed missile of electrons shot out the nearly clear sky, illuminating the land before me like the midday sun. An earthquake-like tremor of thunder followed. Instantly petrified, I yanked up my zipper, tightened my belt like a waistband tourniquet, self-consciously scampered back to Ridgie, and climbed back into the driver's seat. James Taylor's "Fire and Rain" was playing. I burst into hysterical laughter.

All day long, the atmosphere had been flashing its sense of humor. The sky was teasing me, reminding me that, in the end, life has a way of working itself out. Like the atmosphere, the human condition is a product of rules and randomness. Sometimes convention could be tossed out the window. Once in a while, fire could furnish rain.

Journey to the Arctic

Early August arrived before I knew it. I drove back from my internship in Atlanta, renewed my driver's license, applied for a Bolivian visa, and packed my suitcase for the Arctic. I also piled clothing into a second suitcase—I'd be studying abroad in Vietnam, Morocco, and Bolivia during the fall semester of senior year, which would commence just thirty-six hours after my return.

Santa must be around here somewhere, I thought on August 2 as I disembarked an Alaska Airlines Boeing 737 in Nome. I had never been so far north. Located just outside the Arctic Circle on Alaska's Seward Peninsula, Nome is home to about 3,600 people. It's inhabited by a mixture of Inupiat Eskimos and non-Natives, with an industry tied to fishing, shipping, and the withering gold rush of yesteryear.

The late summer air was crisp as I descended onto the tarmac. It was close to sixty degrees outside. The sun warmed my neck but the breeze had an icy tincture to it. Members of the ground crew marshaled us into what might as well have been the Nome Airport, Tire Center, Post Office, and Butcher. Baggage claim was just a hole in a wall that workers slid suitcases and rugged plastic containers through. A few containers of freeze-dried meat could be found as well.

Bob was the expedition's lead scientist, which meant he was in charge of logistics, but a dozen other researchers would be participating in the cruise, too. Each brought with them a graduate assistant or two. At twenty, I was by far the youngest.

We had about a day and a half before we'd be shipping out on the *Healy*, which was on its way up the coast from Seattle. That gave the other graduate students and me ample time to explore Nome. Even though it'd be easy to walk from one side of the city to the other in barely twenty minutes, there were a number of hidden gems—the Polar Bar was a local watering hole, and the Bering Sea Bar and Grille was pretty good, too. We visited the supermarket, where a watermelon cost $18. By some miracle, Nome had a Subway.

USCGC *Healy* is the United States' largest icebreaker, coming in at a whopping 420 feet long. It's designed as a floating laboratory, with ample counter space and room for up to fifty-one scientists. And the Healy is fast, too—driven by 46,000 horsepower, it can cruise at 20 mph.

We boarded the ship via small transfer boats eight at a time on August 4. I still only knew Bob, but my shyness was wearing off. I was among a gaggle of other nerds; we quickly found plenty to geek out about.

Leah was in the process of earning a PhD and was Bob's right-hand person. She had spent years in Oklahoma City, and it didn't take me long to start yapping about my time there. She introduced me to Jessie, an atmospheric scientist at Colorado State University, and her students Julio and Nadia.

Unfortunately, there wasn't room for any more volunteers on Jessie's project, which focused on learning about the role of lofted salt aerosols from ocean spray in nucleating low-level cloud cover. Instead, I'd be aiding Evie, a research assistant who worked full-time at Woods Hole, in collecting and preserving plankton samples. Her team's project was to investigate harmful algae blooms.

We were handed pagers and taught to don mustang suits in case of a shipwreck. Each of us was expected to beep in twice a day to confirm we were still onboard—it was a way to take attendance and a last resort of safety.

—~—

The Earth is tilted at 23.5 degrees on its axis. That's what gives us seasons. When the Northern Hemisphere is leaning toward the sun, we experience summer. When we tilt away, winter.

That 23.5 degree tilt also results in something called polar day and polar night. In the summertime, regions within roughly 23.5 degrees latitude of the North Pole (i.e., in the Arctic Circle north of 66.3 degrees latitude) experience at least one twenty-four-hour period without darkness. The sun never sets, and they're never shadowed by Earth. It becomes the land of the midnight sun.

In the wintertime, the opposite occurs. Polar night caves in over the tundra landscape, darkness collapsing south as the sun refuses to breach the horizon for days or weeks on end. In Kotzebue, Alaska, essentially in the Arctic Circle, the sun is only up for an hour and forty-one minutes on the winter solstice. In Utqiagvik, Alaska, the United States' northernmost town, the sun doesn't rise at all between November 19 and January 21.

Nome isn't in the Arctic Circle, but it didn't take us long to get there. Three days into the trip, I was leaning against a railing at the stern of the ship enjoying the sunset, a sweatshirt all that was necessary to brave the fresh marine chill. Instead of dipping below the horizon, however, the glowing apricot began ascending again like a jetliner performing a takeoff/go-around. It was then I remembered: we were in the Arctic Circle.

The ceaseless daylight made working the midnight shift easier. Evie and I alternated twelve-hour stints of processing seawater in a manifold, a four-pronged apparatus that vacuums liquid from beakers through finely grated filters. Then we folded the sifted remains of plankton and biomass into test tubes, which were frozen at minus eighty degrees Fahrenheit.

My shifts began at 11:00 P.M. and ended at 11:00 A.M. The graveyard shift hours weren't difficult to adjust to—day and night were indistinguishable, making for an easy acclimatization. Meals were still offered in the mess hall every four hours. Instead of lunch, we had "midrats," or midnight rations.

During busier stretches, seawater samples were brought onboard via a Conductivity, Temperature, and Depth (CTD) scan every hour or so. CTDs are akin to reverse weather balloons—they're weighted devices that trap seawater in twenty-four-liter PVC tubes at equally spaced intervals after being tossed to the bottom of the ocean to profile it. In some parts of the Chukchi Sea well north of Alaska's North Slope, the seafloor may be nearly a mile deep. That made for lots of data.

Our instruments allowed us to garner an accurate view of where water masses at various layers of the ocean were coming from. The surface was characterized by a thin pocket of near-freezing water that was low in salinity—it was Arctic meltwater, emanating from liquefied chunks of ice. Human-induced climate change, which has eviscerated much of the Arctic's stock of multiyear ice, is catalyzing the spread of meltwater over the surface of the Arctic and the Chukchi Sea. Beneath that, other saltier water masses, like Pacific subarctic intermediate water or dense circumpolar deep water, could be found.

Seawater is about 3.4 or 3.5 percent salt. That may not sound like much, but that means about seventy-seven pounds of salt are dissolved in every generic cubic meter of ocean water. That's about forty-seven years' worth of consumption by an average human.

There was no Wi-Fi or cell service—we were incommunicado for three weeks. I relished it. Amid slower stretches, the other graduate students and I played 3:00 A.M. board games, a memory I will always treasure. I felt a sense of belonging and kinship that had yet to sprout at

Harvard. The group introduced me to *Oregon Trail*, a game that simulates the westward trek of nineteenth-century pioneers. Christina, determined to win, drew the DYSENTERY card three times; we could hardly contain our hysterics.

The media room had a couple of TV screens and a black couch with cushy upholstering. Anyone who sat in it immediately sank half a foot as the cushions swallowed their body. I knew that if I didn't hold on, I'd sink all the way to a new dimension; maybe I'd end up in Narnia.

One of the screens was connected to a Nintendo. Calder, a student at the University of Alaska Fairbanks, tried to teach me to play *Super Smash Brothers*, but the only game I knew was *Mario Kart*. The other monitor displayed a live webcam view from the *Healy*'s bow. A popular pastime was watching for icebergs on-screen and counting down the seconds until the room shuttered with an audible *crunch*.

I was sitting in the media room on a Tuesday about two weeks into the expedition when I noticed a strange iris-like circle on the webcam. The sun was shining, but a hazy white murk was present outside. At first I thought it was a lens flare, but as the ship turned I realized the arcing band of light was moving, too. I leapt to my feet and sprinted down the hallway, carefully opening and closing each watertight hatchway as I climbed to the top deck.

"Wow!" I shrieked into the uninhabited forsaken landscape. "No way!" A ghostly white fogbow, akin to a rainbow made out of fog, wrapped around the ship. It was pale and monochromatic. Since we were so high up, the entire 360 degree circle was visible. My mind wandered to my bucket list, which I had maintained since grade school.

Warm air advection had brought mild air northward, with temperatures in the mid-thirties. The ice-covered ocean was cooler, condensing the moisture out of the air. That meant freezing fog, which deposited a layer of rime ice on the ship. Supercooled water droplets were present in the air, but their diminutive size meant they were too small to split white light into competent colors through refraction, or bending. Instead, only an acute reddish tinge was visible on the outside. It was like getting halfway to a rainbow.

In the middle of the apparition stood a glory, or a halo-like feature around my shadow. It was caused by diffraction and the propagation of surface waves along individual droplets. It looked like an iridescent halo, a flamboyant bullseye.

A secondary upside-down halation arced upward from the glory, opening concavely toward the sky. It was a reflected light bow—in other words, the sun was shining so brightly onto the placid waters that its reflection had generated a second bow. My camera was clicking nonstop.

—

We returned to shore on August 24, 2018. It was eleven days after my twenty-first birthday, but that didn't matter; members of the science part were determined to commemorate my coming of age. After venturing back onto land, we went to the Board of Trade Saloon. It was my first time having a drink in the United States (I'd had a glass of wine legally in Israel a year and a half earlier). I had no idea what to expect.

Once word of my birthday spread throughout the science party, weather-themed drinks started showing up at my table: a tequila sunrise, a dark and stormy, something else on the rocks . . . three drinks in, I had no idea what was going on. *I don't even remember why I am stressed anymore*, I thought, smiling like a buffoon. *This stuff is great.*

At 4:00 P.M., we returned to Nome's airport. Other than a half dozen random travelers, the thirtysomething of us were the only ones on the flight. I steeled myself as I tiptoed through airports security, hoping TSA wouldn't know I was tipsy. The one-person security force apparently wasn't able to discern my blood-alcohol content with their metal-detecting wand.

Twenty-one hours later, I was back on Cape Cod, where I'd settle for a quick one-day turnaround before flying out to San Francisco. I was visiting Michael, a weather-obsessed friend of mine from Harvard who had grown up in Poland and was back in Massachusetts. Once again, another adventure was before me.

Study Abroad

The smell was what got me first. It was the unmistakable odor of unwashed human, an offensive attack on the nostrils of the innocent, the same smell that I caught in waves riding the sticky-floored MBTA Orange Line train in Boston weeks prior. It was a smell I'd expect after a week camping in the wilderness, but we hadn't left yet. We were in San Francisco. This was day one.

I blinked blindly as my bloodshot eyes adjusted to the darkness of the musty, cramped hostel lobby. Shelves of brochures leaned casually against the wall, a knockoff chandelier whimsically strung overhead. Only a third of its mismatched bulbs were illuminated.

A shaggy-haired man clad in tie-dye stood at the front desk. Both he and the candelabra seemed a bit starved for wattage. A dog-eared calendar depicting a mountain was tacked to the cracked wall over his shoulder. Surveying the room, I knew instantly I was in the right place. But I immediately hoped that I was wrong.

Earlier that spring, I had made up my mind to study abroad, electing a program called Climate Change: The Politics of Food, Water, and Energy. The coursework for my special concentration, which primarily consisted of graduate-level classes and doubling up on courses every semester, was mostly complete. I could either skip out on the fall semester of my senior year or I could travel the world on Harvard's dime. I chose the latter.

There were dozens of approved programs listed on Harvard's website. This one, operated by a third-party company, caught my eye since it visited a trio of destinations: Vietnam, Morocco, and Bolivia. The syllabus emphasized government policy, resource management, and economics. With an extensive background in the physics of climate change, I figured learning the politics of it would make me a double-edged sword.

It turns out the company, without the student's knowledge, had decided to alter its curriculum to focus instead on radical climate justice, replacing economics with ecofeminism, climate physics with climate music and poetry, and politics with protest tactics. As I stood in front of my classmates for the first time, the gravity of my situation began to dawn on me.

Every stool, seat, couch cushion, and haphazardly carpeted strip of floor was occupied by eclectic-looking students, the same way a flock of pigeons overtakes a park statue. Some were perched on suitcases, duffle bags, and pastel pillow cases, though two of the pillow cases had people inside them; I realized they were being worn as shirts. The words F*CK MONSANTO were scrawled in permanent marker on one of them.

Every backpack had carabiners dangling from them like makeshift Christmas ornaments, but I knew what I had gotten myself into wasn't about to be a holiday. Each hardtop suitcase was adorned with bumper stickers, reminding me of the back of the station wagon of my neighbor Anne-Marie, a self proclaimed spiritual healer.

About two thirds of the gaggle of college students convened in the suffocating lobby sat silently, some staring into the middle distance with a glazed look in their eyes. Others compared rubber wristbands, each bracelet accessorizing the wearers with the causes they so bravely supported. Two barefoot girls—one short with curly hair, the other tall and sunflower-like—were playing a game of patty-cake. The group looked to be only a quarter male.

"You must be Matthew," whisper sang a voice from behind me, mellifluously poetic yet inexplicably devoid of sincerity. I pivoted on my heel, slipping my backpack off effortlessly and resting it atop my unwieldy suitcase. I glanced up, finding myself face-to-face with Heather.

Heather was the "traveling fellow." In essence an on-site counselor, she was responsible for facilitating country-to-country transitions during our semester abroad. I had video chatted with her weeks earlier. It was an interaction like none other I had experienced. During the call, she wove a tale of her life story—being born into a mixed-class background (only one side of the family was rich) in Brattleboro, Vermont, working as a cross-country ski instructor, creating her own racial justice and awareness organization, and serving as a volunteer carpenter in poverty-stricken areas of the Appalachian Mountains. She had just turned twenty-six, and, while I was curious, I never asked why she no longer held any of those unique positions. I was about to meet my answer.

"Ha, ha, ha," Heather said, not laughing, but rather forcibly pronouncing the syllables of canned laughter. She wore purple sandals, green cargo capri pants that wrapped halfway down her shins just above her ankles, and a loose-fitting men's black T-shirt with the word OCCUPY emblazoned in yellow. A red handkerchief hung from her back right pocket; I remembered from the video call that she ask we pay close attention to the color each day.

A subdued grin emerged, wide but seemingly rehearsed. An overgrown buzz cut rested atop her slender face, which remained contorted in a seemingly involuntary smile.

"So, how would you like me to greet you?" Heather asked, her head tilting to one side. She appeared genuinely puzzled. So was I.

"I'm sorry?" I answered, my right hand resting atop the handle to my roller suitcase as I stood there awkwardly. Sweat was beginning to soak through the back of my aqua-blue polo shirt. The lobby was roasting. I had flown to San Francisco wearing that and khakis, hoping to make a good impression on my peers. I wanted to dress professionally for day one of my climate policy and government program.

"Well, we could bow, high five, fist bump, or hug. I prefer hug, but you might not be a hugging person," Heather said softly, leaning in with a blank expression on her face. I got the sense that my answer would seal my fate.

"Um . . . sure," I said, my eyebrows leaping as I smiled politely but uncomfortably. I clenched my jaw gently while raising my arms slowly, as if hoisting sails to catch a tailwind that would whisk me away. Instead, she spoke again—monotone and quiet at first, but then cautiously eager.

"May I ask consent to give you a hug?" she asked, once again speaking slowly and smoothly. It was the same tone I used on baby animals.

"Uh, yeah," I replied, instantly admonishing myself. *Uh*? Had I really said that? It was the second filler word in five minutes.

"Great," Heather confirmed, as if announcing the time. She stretched out her arms, a dozen or so Technicolor rubber bracelets gliding down from her raised wrists. I knew she believed herself to be a warrior who had waged many valiant battles. She noticed me eyeing her bracelets. She grinned as her eyes widened.

"This reminds me of a song I created," she said.

By midafternoon, I had convinced myself that substance abuse was on my horizon. I had 106 days until the program concluded; five hours into it, I was already counting down the hours.

"We're going to begin with silence," Heather said, tiptoeing in between thirty-five other students and myself as we clustered on the thirteenth floor of a passé Bay Area high-rise. The conference room had vaulted ceilings from which specs of peeling paint dangled as if barely holding on. Metal radiators spanned the length of the wall adjacent to the windows. They looked like they had been there since the 1906 earthquake.

"The quieter we are, the more we can hear and reflect on the voices of our ancestors," she explained softly. "Who are we? Let's introduce ourselves to each other in silence."

I looked around incredulously. The other students were nodding. Some were tapping their feet, while others thumbed absentmindedly at their wristbands. A few snapped after each instruction Heather spoke; I gathered that they were trying to communicate agreement.

"We're going to wander—not in any direction, but just wander, like travelers; we're guests in this space," she explained, waving her thin arms like limp spaghetti. "When I say 'freeze,' turn and stare into the eyes of the person closest to you."

Silence flooded the room as I searched around for the hidden cameras. I was definitely being Punk'd. Only the periodic echo of shoes scuffing against the warped wood floor could be heard. I stare at an entanglement of three dozen zombie-like people weaving around the room. I felt like I was watching a video of an antisocial cocktail party on mute.

"Freeze," Heather said delicately. The mass of students froze in their tracks. Was this the woke version of musical chairs? Before I could ponder that thought, a tall, lanky girl with a name tag reading HELLO! MY NAME IS SUMMER-SOLSTICE pivoted toward me. She peered into my eyes with a blank, empty expression. I was surrounded by sleepwalkers.

The exercise went on for another twenty minutes. At one point, I lost a staring contest to a man named Ron who apparently was our traveling professor. He looked to be about sixty years old, with glasses and fraying silver hair cascading down to his shoulders. He wore sandals, white canvas pants, and a V-neck T-shirt that read PEACE on it. Tufts of matted chest hair spilled over the collar.

When the icebreaker was over, he shuffled over next to Heather. He stared at the ground, his arms remaining still as he walked.

"I'm Ron," he croaked nervously to the group, his voice shaking. "I'm a sociologist, but more importantly, I'm an activist. Climate change is the product of capitalism, and we can only fix one by abolishing the other. I am thrilled to be teaching two classes this semester on the radical social changes we need to end climate change and get justice."

"Mmmmmmmmm," the group replied in unison, bobbing their heads and snapping. I was bewildered. I thought I was here to learn governance and policy.

Just then, Fern, a program coordinator who was overseeing the launch of our group in San Francisco, floated into the room, carrying a half dozen stacked aluminums trays.

"Our nourishment is here," she sang. She traipsed across the floor barefoot, gingerly placing the tins on a plastic folding table before peeling back the cardboard. Steamed beets, yams, and quinoa. All program-organized meals in San Francisco would be vegan.

Things didn't get any less weird in the week ahead. Panic began to set in. *Do I drop out?* I thought. *Go back to Harvard and get a masters in economics? Did I choose the wrong career path?* It was senior year, and while all my Harvard friends were being recruited for consulting, finance, and tech jobs in shiny city skyscrapers with six-figure salaries, I was singing about organic vegetables in San Francisco. How was I going to complete my major with this? My new colleagues certainly took some getting used to.

We spent our days inside the sunlit second floor of the Women's Building on 18th Street. It was a community center whose exterior was painted with stunningly vibrant murals, and large picture windows allowed fresh air and daylight to breeze through the building. The scene was bright;

I was a dark cloud in valley of flickering rainbows and llamas masquerading as unicorns.

Our final day in San Francisco was spent determining "where we belong." Kevin, our guest speaker, explained that "our people" are those among whom we share something that "others" us. He distributed worksheets as we sat cross-legged on the floor.

I glanced down at the paper. Ten empty boxes were printed on it, each headlined with categories of things that might "set us apart": the first box was labeled SEXUAL ORIENTATION. *Not today*, I thought. I returned to typing a newspaper article on my laptop, ready to pull up my pretend NOTES tab if Heather tiptoed by.

"If you're done already, you can reflect on your balance between being othered and being privileged," Heather said, appearing in front of me like a genie. *Maybe she could smell fear*, I thought.

"Would you like to share your boxes?" she asked.

"No," I replied. Everyone else was still writing, but a few students turned and stared at me. "Who I am doesn't come from boxes or categories. That seems counterproductive. 'My people' can be anybody." She thought for a moment.

"We're going to have to have a conversation with Zach about your laptop usage in class," she said as she took another bite of granola.

~

At long last, September 12 rolled around. It was the point of no return. We were slated to fly from San Francisco to Taipei before connecting to Hanoi, Vietnam. After "classes," I walked to Walgreens to stock up on table salt, my number one travel essential.

At 7:00 P.M., we loaded our suitcases into a chartered coach bus that would take us to the airport. Francine was our "travel POD," or the Person of the Day, which meant she would be facilitating our travel. After everyone had settled into their seats, she read a poem for the group. The bus lumbered down Highway 101.

At least I'm not in this alone, I thought. I had acquired two close friends in San Francisco, Jason and Kelby, who were equally dumbstruck by the program thus far. They too had thought this would be a course focused on climate change and real social policy that would help people who needed it. I rejoiced when I found out they had only been faking it to remain well-liked by their peers. Behind the scenes, they were just as fed up as I was.

Jason was a junior at the University of Richmond studying economics, political science, and leadership. Originally from Metuchen, New Jersey, he fit the mold of a classic self-absorbed fraternity guy, but if you got him talking about global climate economics, his brash persona was suddenly replaced by wide-eyed enthusiasm. Sometimes he concentrated so much on his explanations that he would forget to blink.

Kelby was the polar opposite. While Jason was a city dweller, Kelby came from rural northeast Oklahoma. We became fast friends during the first week of the program as soon as I learned where she was from.

"Oklahoma," she had told me originally, declining to go into more detail.

"Where in Oklahoma?" I responded.

"You won't know it. It's in northeast Oklahoma," she said.

"Try me."

"Pryor Creek," she said, shaking her head.

"Oh, you mean off the Claremore exit along Highway 20 past the QuikTrip?" I replied. She was stunned, if not maybe a bit suspicious. I quickly explained that I was a storm chaser, and her turf became my stomping grounds each spring during severe weather season. Plus I loved Pryor Creek for its Arby's.

As I learned more about her story, I couldn't help but be amazed. Born to teen parents, Kelby grew up in a mobile home in the Cherokee Nation.

Her father was out of the picture, and her mother had struggled to provide for the family early on. She was raised in part by her grandparents.

The economy of Mayes County, Oklahoma, didn't exactly afford many opportunities to advance in life; earning a steady income often meant working long hours at multiple jobs. The same was true with the education system. Some of the worn-out textbooks in the county's underfunded schools referred to the Civil War as the War of Northern Aggression, and largely glossed over slavery.

Kelby had myriad disadvantages, yet she excelled in school, became a decorated recipient of national scholarships, and earned a full ride to Duke University. (Nowadays she's wrapping up law school in Colorado and working on a book.)

It wasn't until I met her mother, Tammie, in 2019, that I discovered one of the reasons for Kelby's success. Tammie loved her kids more than anything else in the world. Within five minutes of meeting her, it became clear to me that Tammie would do anything for them. That meant helping Kelby get ahead in the world through hard work, grit, and determination.

Jason sat next to me on the bus to the airport; Kelby was somewhere toward the back.

"Where are you sitting?" he asked.

"31C," I replied with a smirk. Heather had emailed the group's itinerary to everyone; I logged in and changed my seat so I wasn't within the group block. I couldn't risk being slingshot through the upper atmosphere at 500 mph with someone next to me singing.

"You didn't move me?" he asked. I chuckled.

"I actually put you on a flight to Tahiti," I replied, rolling and then narrowing my eyes.

"I wish," he grumbled.

—~—

"Capitalism!" Jerusha screamed. A pair of flight attendants deviated around the group as we clogged the hallway in front of the airport entrance. The bus had just pulled away from the curb, and Heather was standing in the doorway as if preparing for a marauding lunge. The automatic doors continuously opened and closed on her bag.

"Is!" Riley shrieked. A family walking by tugged their young children closer. The parents glowered at us.

"The!" Jessie bellowed vapidly. Merchants at a nearby souvenir kiosk eyed one another.

"Worst!" Juniper echoed. I cringed

"BUFFOONERY!" Francine sang out.

"To!" yelled Miley. Then a pause followed.

"Ever," Jason said calmly. I looked over at him. He caught my eye and shook his head.

"EXIST!" Summer-Solstice concluded, standing on her tiptoes as she stared triumphantly into the middle distance. I quietly snuck a few paces away from the group. *Maybe if I stare at my phone, I can pretend not to know them.* This was how we took attendance. It was Heather's idea.

Checking in to our flight was an equally embarrassing spectacle that attracted attention far and wide within the terminal. Every four or five minutes, Heather yelled "Mic check!," to which the group would parrot back "mic check!" as a signal that they were listening. Most of her announcements pertained to poems and protest songs that she perceived as relevant. It didn't take long before we were bellowing melodies in TSA, though they sounded more like dark incantations.

We finally made our way to the gate by 9:00 P.M. Jason and I walked alongside Damien, chatting about our previous international escapades. Damien was a junior studying political sciences at the University of Pennsylvania; unlike Jason, however, he was tough to read. Behind closed doors he complained incessantly about the program, but in Heather's presence was an enthusiastic sycophant. I couldn't tell which was the real him.

"We'll meet here at 10:00 P.M. for closing POD," Heather said without blinking. She was speaking to the group, but alternating her stern gaze between Jason and me. That gave us just under an hour to grab dinner, and boarding wasn't until 11:25 P.M.

Jason and I strolled through airport, carefully keeping track of the time. Kelby decided to look for food with Karen, Damien, and Allie; she was unwittingly tagging along a trio embroiled in a contentious love triangle.

We all returned at roughly the same time, rejoining the group at the gate around 9:52 P.M. Heather greeted Kelby, Karen, Damien, and Allie before strolling over to Jason and me.

"You're late," she said. She stared at us with a cold, hardened expression.

"Heather, it's 9:53 P.M.," I stated neutrally, my voice unwavering. "We're actually seven minutes early."

"You're late," she repeated, nodding and leaning toward us. Jason and I said nothing. Heather appeared enraged. "You were supposed to be here at 9:45."

Kelby, overhearing Heather, turned and shot me a confused glance. I flicked my eyebrows as if to say *I don't know, either.* I turned to the remainder of the group.

"Guys, what time were we supposed to be back?" I asked smugly.

"Ten," said a few. Heather tightened her jaw. I was sure smoke was about to pour out of her pointed, elf-like ears.

"You're late," she said again, as if to convince herself. "If you're late again, you're going to lose your privileges, and we'll have to have a conversation with Zach." Zach was the program director whom we met in San Francisco. Before I could say anything, she stormed off.

An hour later, the thirty-eight of us lined up in the jet bridge, filing one by one into an EVA Airlines Airbus A350 plastered with Hello Kitty decals. Jason found it hilarious.

“The engines are powered by song,” he said. I laughed. Just then, I felt a presence behind me. Heather was nearby.

“Matthew, I wanted to approach you about your responses earlier,” I heard. I turned around. Heather was barely three feet away and was talking to me, but she wasn’t looking at me—her eyes were darting back and forth. I said nothing, curious (and slightly amused) to see where this would go.

“I didn’t need a correction earlier,” she said, enunciating her words carefully like a contestant on *Wheel of Fortune*. “I can’t have that again.”

“I’ll be sure to abide by whatever you say, Heather,” I said, smiling. “And that means if you say ten, I’ll be there by ten, guaranteed.” I grinned broadly, feigning enthusiasm.

“I need more ‘yesses’ and less ‘nos,’” she replied, demanding an apology. “You also didn’t give your eye contact to the Person of the Day, and you didn’t actively participate in our gratifications or song. Traveling is stressful, and I’m trying to make it as easy on everyone as possible.”

Two hours into the fourteen-hour flight across the Pacific, I discovered what does make traveling as a group easier: sauvignon blanc. We would become fast friends that semester.

Riding Away

I don't think my mother would like this," I said to Jason and Kelby. At least I had a helmet. Besides, for $6, I couldn't pass up a bargain this good. Back home, I'd be paying fifteen times that.

It was early October, and we had been in Vietnam for a month. The prior four weeks had been filled with pure insanity. Now I was escaping, literally and metaphorically, via motorcycle.

It was the first time I had ever truly throw caution to the wind, but I needed an out. I had been punished for being "speciesist" after making a "why did the chicken cross the road?" joke, was told that my facts were "irrelevant in a discussion on feelings," had my laptop taken away for "not being actively engaged in group song," and was informed that the amount of table salt I was lugging around was "excessive." Even Phoung, our local country coordinator contracted by the study abroad company, thought Heather was loony. And she had it in for me.

—w—

Our first stop in Vietnam was Hanoi, a city of five million situated in the northern part of the country nestled within the Red River Delta. It was a charming amalgamation of cultures that at times seemed contradictory; the shades of Western influence were perceptible everywhere, but to me it was an aftertaste after soaking up all the traditional Old World temples and architecture.

Commercialized convenience stores flanked unregulated street markets where elderly women sold fruit and children peddled cell phone chargers, adapters, and electronic merchandise; people of all ages knew how to haggle. Counterfeit name-brand clothing hung from sheet metal awnings propped out from buildings, with mounds of "Norht Face" jackets and "Adiddas" athletic gear piled on dirt sidewalks or tile patios. The streets were overflowing with stuff.

I was fascinated by the economics of the region. It seemed that everyone was poor, yet no one was destitute. Most households took in only a few hundred dollars per month. That wasn't enough for laptops, tablets, or vehicles, yet it was sufficient for a family to scrape by on. Food was cheap.

One was never more than fifteen feet away from a cauldron-like vat of pho, Vietnam's signature soup, which could be found cooking on every street corner. Savory scents of beef broth and cilantro wafted through the air. Fifty cents would get you a hearty portion, usually scooped into a bowl by a grandmotherly woman wielding a ladle. It was best enjoyed squatting on plastic stools alongside a few close friends; pho was a sort of communion, a meal of trust central to Vietnam's rich culture.

The biggest shock came on the roadways. Prior to our arrival, we had been told "don't wait; cross slowly and don't stop." Within ten minutes of landing, it became apparent why. Motorcycles and scooters packed the roadways and swarmed like angry bees, weaving chaotically around one another and occasionally hopping over curbs. The air was abuzz with cartoon horns.

Road laws seemed virtually nonexistent. There were signs and traffic lights and lines painted on the ground, but they were mere suggestions. If the shortest distance between two points was a straight line, that was the path to be taken, even if it meant going against the flow of traffic or hurtling down a sidewalk.

I had never seen so many motorcycles clogging a roadway, nor witnessed them used for transporting crowds and goods. Entire families of four could fit on a scooter with backpacks and bundles of groceries. Some mopeds carried cartons of fruits and vegetables or drums of water, while others were stacked with chairs, appliances, couches, caged chickens, and burlap sacks of cabbage. They reminded me of something out of Dr. Seuss's *The Cat in the Hat*. Crossing the street, I couldn't help but feel like a stone in a river; I silently prayed the traffic would bifurcate around me.

By the third week in Vietnam, we had relocated from Hanoi to Da Nang, a seaside city about 250 miles to the south. If it hadn't been for the program, Da Nang would have been paradise. The warm, tropical waters of the East Vietnam Sea were delimited by picturesque sandy beaches, with French-inspired mountain resorts in the Bà Nà hills just an hour's drive inland. The city's airport was located downtown, and our homestays were a short jaunt away.

Instead of Uber, we utilized a service called Grab, which consisted of motorcycle taxis that arrived at a moment's notice. By day, I snuck my blue Camelback water bottle to class filled with inexpensive Vang Dalat white wine; it helped me tolerate the hours of song. After classes, my friends and I frequented local bars; I always kept my laptop handy to write *Washington Post* articles for morning publication stateside. At night, residents and tourists alike flocked to the Dragon Bridge over the Han River, which breathed a flame of fire shortly after each sunset.

While intellectually stimulating dialogues were few and far between in the classroom, there were a couple takeaways buried within the noise. Among them was Vietnam's involvement in Reducing Emissions from Deforestation and Forest Degradation (REDD+). REDD+ is a program first negotiated by the United Nations in 2005. It aims to slow the pace of climate change through, among other things, improved forest management.

It's believed that deforestation and forest degradation could account for more than a fifth of net greenhouse gas effects; vegetation is integral in carbon sequestration, so preserving it is a must. Vietnam adopted REDD+ in 2009, with forty-nine participants, primarily Global South nations, enrolled by 2014. The framework creates financial incentives for maintaining "stocks" of carbon contained in trees, as well as to guide long-term strategy toward sustainable development.

Our time in Dan Nang was interrupted by a four-day stay in Hoi An, a mellow town about an hour to the south known for its world-class tailors. Jason suggested we visit one, so we did. A well-dressed, flirtatious woman named Ruby sat in a musty workshop surrounded by heaps of sample fabrics; she talked me into ordering a couple custom suits, four shirts, and a pair of handmade shoes. The bill for everything came to just under $300, and everything was ready the next morning. I was dumbstruck by the results—the craftsmanship was impeccable, and the quality of the fit exquisite. (To this day they're the only suits I'll wear on television, and I'm planning a trip back to see Ruby and purchase more someday.)

Despite the relaxed nature of the resort town, escaping the throes of Heather was a Sisyphean task. Everywhere we turned, she was there; I could only fill my water bottle with wine so many times before its escapist powers lost their potency. Finally, Jason, Kelby, and I had enough.

—~—

"She's going to know," Jason said. "We can't all fake being sick. She knows we're always hanging around each other."

"Yeah," Kelby said, nodding in agreement. "Should we use our program pass?" I rolled my eyes. So many people were feigning illness to escape Heather's startlingly intimate group bonding exercises that she instituted a one-skip policy per country to maintain her illusion of control; she called it the "program pass." We decided it was our best option.

"I'll book them," Jason said, whom, by then, I had nicknamed "YelpMom" thanks to his obsession with Yelp and TripAdvisor reviews; no restaurant nor café could be eaten at without a careful examination of online ratings. Anything below 4.2 stars was a no-go. He'd sooner starve than eat at a three-star establishment.

Despite Kelby's and my taunts, Jason had a knack for finding and booking unusual activities off the beaten path. We came to trust him blindly, even if that meant hopping into random vans, wandering down alleyways, or navigating the catacombs of a basement market. He always knew where he was going. Ordinarily a control freak, I learned to just go along for the ride.

The motorcycles only cost $6 to rent for the day, including gas, with an extra dollar for delivery to our hotel. We didn't think that part through—blatantly disregarding the rules was a lot easier when you weren't doing it thirty feet from where Heather was staying. This wasn't exactly a clandestine operation.

When the motorcycles arrived, I mounted mine tepidly, eyeing it like I would a wild animal. I pictured a mechanical bull like the kind found at country and western bars.

I revved the throttle and was immediately jolted back by the speedy acceleration; this thing had some power. I pressed my legs inward and hunched down toward the handlebars, tightening my grip while making myself more aerodynamic. Jason and Kelby followed suit.

"Later!" I yelled, barely audible over the engine before zipping off. It sounded like an angry chainsaw.

We rendezvoused a half mile up the road and discussed our options, settling on a beach thirty miles down the coast as our destination. The journey started a bit shaky and unsteady, but we got the hang of it after a few minutes; before long were weaving in and out of lanes in true Vietnamese fashion.

Unlike the congested arteries of Hanoi, traffic dropped off markedly as we reached the southern outskirts of Hoi An. By the time we rounded a corner from HL 15 onto a highway labeled 129, we were the only ones in sight.

The smooth, paved road ran straight as far as the eye could see. Sandy-colored fields, shrubs, and sporadic palm trees ruled the landscape, with intermittent clumps of tin-roofed shacks clustered around occasional intersections. Most structures had banners depicting the Coca-Cola or Lays Chips logos draped from their shutters. I clasped the throttle more intently, twisting it back toward me.

The thin needle on the speedometer smoothly glided to the right like the hands of an analog clock. Fifty, sixty, seventy, eighty . . . I could see Jason appear smaller in the handlebar-mounted mirror; Kelby, trailing behind, was a tiny speck. I shook my head; if Jason was a gentleman, he'd be behind Kelby.

Even though it was eighty-five degrees outside and oppressively humid, the air blasting by my face felt cool. *It's the same premise as wind chill*, I thought. Humans constantly radiate heat, generating a small insulating cushion of warmth around them. If one moves fast enough, they outrun that invisible layer and feel cooler. (It also has to do with wet bulb physics; air moving swiftly past someone will parch the sweat on their skin more efficiently, resulting in additional evaporative cooling.)

I accelerated faster, the melodious hum of the engine blending with the sputtering hiss of air deflecting around me. The red needle teetered closer to 100 kilometers per hour; it was taunting me. Narrowing my eyes, I surveyed the horizon for any signs of imperfections in the road: no potholes, stones, or divots. Once more I doubled down, my pulse quickening as the speedometer spiked into the triple digits. I was free.

For the first time in months, I felt a wave of bliss wash over me—apparently, at the equivalent of 62.5 mph, I could outrun whatever was on my mind.

I forgot about the lunacy of the program; the building anxiety I held over my post-graduation plans abruptly melted away. For a few brief moments, any self-doubt I carried over whether or not I was embarking on the right career path vanished. The pervasive loneliness that had been a constant during my prior three years of college now seemed a distant memory.

It occurred to me that I suddenly wasn't lonely anymore; I was flying down a freeway on the opposite side of the world with two friends I was sure I'd keep for life. They already knew more about me than anyone else, and yet their discovery hadn't changed anything. I had a feeling they were sticking around.

—w—

Is that a roller coaster? I asked myself. We'd been passing through open fields and nothingness for nearly half an hour. Now we were driving next to a misplaced retail oasis.

I slowed down, my earlier exhilaration replaced by a hollow optimism. Lush, meticulously maintained lawns surrounded teal-blue swimming pools, a playground-like water park towering four stories high. Bright neon beach umbrellas stood obediently, spaced evenly on a cobblestone patio a few hundred feet away; a lazy river snaked around a spattering of amusement park rides, erected next to a row of Western-style buildings designed to look like a typical American Main Street. A pair of hotel high-rises, pylons of opulence, stood boastingly in the distance.

I stretched out my left arm to signal an upcoming turn before leaning into the curve. Jason and Kelby were close behind. I spotted a sign that read VINPEARL. From what I gathered, the beachside resort was brand new and empty.

I eyed the sky—it was early afternoon, and tufts of puffball cumulus clouds were sprouting. Monsoon season ran year-round in Vietnam. Northern Vietnam, including Hanoi, is prone to daily downpours during the late summer, but most of Da Nang's occur between October and April.

After another ten minutes of driving, we finally turned onto a dirt road. Evidently, Google Maps is very generous in what they consider a road. A two-lane dirt strip turned into a one-lane grass pathway, which narrowed to a worn-down and overgrown pedestrian walkway barely a meter wide. I lead the way on my motorcycle, aware that I couldn't stop; if I did, I'd tip over.

Palm trees shaded the tunnel-like footpath as the hard, earthen surface became sandier. Eventually, our tires began to sink into the ground. At long last, I spied a few shelters made of plywood and corrugated metal, catching sight of an opening up ahead. We had made it.

I shut off the ignition and hopped off the motorbike, leaning it against a short, rusted chain-link fence. Jason did the same. Then Kelby's engine went quiet.

"Gah!" I heard her shout, a mix of concern and comic relief. I turned around. Her bike had toppled over when she tried to get off it, the kickstand poking several inches deep into the beachy ground. She was stuck beneath it, laughing and flailing her arms.

"Can I get a little help over here?"

Life in the Desert

"To not receive a single thank-you note, an acknowledgment, or an appreciation was hurtful," Heather whispered. Silence gripped the room, the dry Moroccan air gradually thickening with building tension and consternation. Jessie stared at the ground with watering eyes as Miley shuffled uncomfortably in her seat. I was three water bottle "glasses" of chardonnay in and reclined in my chair with my laptop open. My focus was locked on a satellite display of Category 5 Hurricane Michael as it churned toward Florida's Big Bend. Thousands of people were about to lose everything, and we were comforting Heather for feeling her travel poetry was underappreciated.

It was October 10, 2018. We had only been in Morocco for about a week, but Heather had managed to take her madness to new heights. Classes were held in a mosaic- and sandstone-constructed community center in the medina, or main town center of Rabat, a city of roughly half a million on Africa's northwest coast. The afternoon session was supposed to be over forty minutes ago, but Heather was on a tirade. We weren't going anywhere anytime soon.

"My role is to facilitate transitions," Heather said indignantly while clutching her stomach, as if reeling from a sucker punch to the gut. "I spent time recording some soothing chants about our ancestors. Maybe it was just because we were in the airport, but to not hear a single thing from anyone did not reflect our community values."

Squeak! The entire classroom turned to face me. I was tightly gripping a rubber bear that Jason, Kelby, and I had nicknamed "Bruce." Hiding it in each other's bags had become a running joke; on this particular day, I had introduced it to the class as my "emotional support object." If you can't beat them, join them, I figured.

"Did you want to share something, Matthew?" Heather asked sinisterly, incredulous that Bruce would interrupt her. I slouched forward in my seat, snapped the spout of my plastic water bottle into place, and began my soliloquy as if on cue.

"Heather," I said, smiling with cold, empty eyes. "Some of us are functioning adults who don't need a comfort poem or song to board a plane." Miley gasped as Francine eyed me curiously. Ron was chewing on sunflower seeds, but his jaw suddenly went still.

"Perhaps instead of forcibly extracting appreciations from people, you might consider being open to genuine interactions when people actually appreciate something," I said, nodding with raised eyebrows. "I'm pretty blunt. When someone does a good job, I usually say 'good job.'"

I reached down and lifted the plywood panel of the rickety folding desk, promptly collecting my items as I shuffled toward the door in silence. Thirty-six pairs of eyes followed me. I knew that Jason, Kelby, Damien, Allie, Karen, Victoria, Race, and Shelly were all jealous.

"I'm not done," Heather shouted in shock, no longer feigning a jarring state of illusional peacefulness. Her eyelids quivered as she clenched her jaw.

"I am," I said, smiling and bowing my head politely. "To quote Maxine Waters, 'I am reclaiming my time.'"

I stumbled down the steps in a hopscotch fashion, impressed I hadn't slurred a single word. I squinted as my eyes adjusted to the abrupt brightness that radiated outside of the stuffy, dusty classroom. I wandered through street markets until I found what looked like a pizza joint. "Thirty dirhams isn't too bad," I said, plopping down on a red plastic stool. I opened my laptop and continued forecasting for the next forty-five minutes until Jason and Kelby finally arrived.

"That was bullshit," said Jason, huffing and puffing like the Big Bad Wolf. I tilted my head up from my laptop and smiled, amused.

"She talked for eighty minutes about how she felt unappreciated because we didn't like her 'be easy, you're coming home to yourself' song."

"I know," I laughed. "I was there . . . at least for part of it."

"You're going to get hell tomorrow," Kelby said. I shrugged.

We meandered half a mile to the beach, a fiery crimson brimming over the horizon. Caterpillar-like articulated busses lined the roadway, each window pane on them shattered or punctured with small holes. They looked like props from a wartime movie.

It was my first time looking west at the sun setting over the Atlantic. Growing up on Cape Cod, I had only ever seen sunrises over the water. As the peach-colored orb began sinking below the distant waters, I suddenly remembered a piece of mystical meteorology that had been fabled to exist in similar setups.

"Guys, we've got to look for the green flash!" I hollered, startling Kelby and Jason. We were standing on weathered rocks that rose about twenty feet over the splashing waters, planted like sleeping giants guarding the coast. They looked at me quizzically.

"Sometimes the last rays of sunlight escaping over the horizon look green," I explained, unaware if they actually cared or were being polite. "It's kind of like a mirage. Colors other than green or blue are scattered or absorbed. Right before sunset, that green isn't overlapping with red or yellow or anything."

They shrugged, but I knew I had captured their attention—both were squinting and peering at the distant water.

"Ten, nine, eight," I called out. "Seven . . . "

Suddenly, a tinge of lime green appeared at the lip of the horizon, growing brighter and spreading horizontally before vanishing altogether. I blinked hurriedly, hoping the impression was still emblazoned into my pinhole pupils.

"That was it!" Jason said eagerly, visibly excited by what he had seen. Kelby nodded, her eyes widening.

"That was *so* cool!" she echoed.

I smiled. It wasn't much, but it was a moment where I got to share the atmosphere with my friends. I didn't expect to *actually* see it, but I was glad that I had people alongside me to appreciate the experience together. Another item off my bucket list.

"Wait until y'all see the night sky Monday night in the desert," I said, looking ahead to our trip to the Sahara the following evening. "You'll see more stars than you ever knew existed."

I crouched beneath a waterlogged awning, waiting for the rest of the group. Jason and Kelby stood beside me, with Damien, Allie, and Karen lagging behind. We were waiting for Race and Shelly, who were collecting their luggage out of our Airbnb. We'd rented the entire dwelling, but apparently it came with a woman who spent all day in a small, cramped room off the kitchen frying potatoes.

A heavy rain was falling; it had arrived the night before at the same time as the eight of us. The old-style town, Fes, was a stopover point on our way to the Sahara Desert. We had a five-day break in the program, and we sure as heck were going to escape Heather.

Fes, a city of 1.2 million, wraps around a sleepy settlement reminiscent of an Egyptian castle. Modern suburbs sprawl around a maze of brick and sandstone corridors, which is made up of winding alleys and hairpin tunnels. I couldn't tell if I was inside or outside, in a hallway or on a walkway. Street vendors crowded every corner, pressing against the chalky sides of adjacent buildings whenever the garbage donkey strolled through. (There were no trucks in the city center, so a mule was used to collect rubbish.)

A strong cold front was sweeping through the region, trailing to the south of low pressure passing near the Strait of Gibraltar. It was moving to the east, which meant a dismal drive would be ahead of us as we followed the front to Merzouga, a small town in southeastern Morocco on the Algerian border. I shivered. The day before had been in the lower seventies, but the cold rain had extinguished my pilot light. Now we were hardly in the fifties.

"He said he's this way," Damien said, leading us down a sandy cobblestone walkway toward the edge of Fes's medina, where we'd be meeting our tour guides. I wasn't looking forward to the ten-hour drive.

Our guide's name was Youssef. Damien and Karen had been in charge of arranging the tour, and negotiated with Youssef to secure the excursion for $300 per person. That included transportation and ATV rentals. It was a steal. We divided into two groups and climbed into twin four-wheel drive SUVs. Jason, Kelby, Race, and I hopped into the second vehicle, while Damien, Allie, Karen, and Shelly took the lead car. We shed our raincoats as soon as we entered, but it was no use: the windows immediately fogged up.

The driver, Amir, remained silent, accelerating down the flexuous road as we exited Fes. The drone of the defrosters grew louder as the translucent windshield became opaque. Amir grabbed a handkerchief from his pocket and wiped away small a patch of condensate, hunching down to see outside. The windshield wipers flailed wildly as the rain beat down harder. Amir squinted. I felt like I was on an airplane at night; I had no view of my surroundings, nor any spatial awareness.

The smooth, paved roads gave way to cracked strips of tar riddled with potholes. I was glad I had eaten a small breakfast.

"Ruuuuuuurrrrrp!" Race hollered from the front seat, suddenly leaping up, noticeably startled. I glanced toward the rearview mirror mounted to see what the commotion was about, but noticed there wasn't a rearview mirror. Apparently, it had flung off the windshield and, to everyone's surprise, into Race's lap.

"I guess I'll just hold this," he said while erupting into laughter, inhaling air in an uncontrolled series of snorts and giggles. It was contagious. Amir, meanwhile, didn't look up; his knuckles were white as he gripped the steering wheel. It felt about right given the terrain.

The curving road began to climb as the vehicle pitched upward, wind-driven rain now splattering the windshield dead on. The downpour intensified to the point that we were raising our voices to speak. Shrubbery and palm trees were replaced by pine trees poking out of the red, burnt landscape. Some huddled together, as if hiding from the cold. The dashboard displayed a reading of nine degrees Celsius (forty-eight degrees Fahrenheit).

Rolling undulations in the landscape became hills and eventually mountains. The once-gentle swaying of the vehicle evolved into steep uphill lurches and careening downward swerves. I knew we were ascending higher based on a sealed bag of potato chips resting on my thigh; ambient air pressure was falling, leaving the constant volume of trapped air to puff out the bag. Even snacks can act as a makeshift barometer.

I looked out the rear windshield; through the mist, I could tell we were along the rim of a valley. Only the landscape below was visible beneath the curtains of fog. We were disappearing into the clouds.

Swoosh! A small speck of white sliced past my window. I wiped away more of the fog. Another passed, and then another, sailing through the air like feathers glissading whimsically to the ground. One stuck to the glass, its rimy edges accreting before merging with a drop of water and sliding downward.

"Hey, Capooch, it's snowing," Jason said, tapping me on the shoulder from the rearmost seat. I leaned forward, transfixed. The sky had surprised us with a cloudburst of confetti.

Five minutes later, the burgeoning blitz was a raging blizzard. Visibility dropped to near zero as we chugged along through elevated plains in the High Atlas Mountains. Our progress slowed; 40 mph to 30 mph to 20 mph; eventually, we were moving along at a crawl.

"Ope, ope, ope!" Kelby shouted. I reached for the pulldown ceiling handle. We were sliding diagonally, but Amir had the wheel pointed straight. All I could see were brake lights from the lead vehicle in our stunted, intrepid caravan.

Thump. The right front tire exited the roadway as we glided toward a ditch like a wrong-way toboggan. Amir cranked the wheel left, the back of the vehicle whipping around like a pendulum as we skidded over rocks and clumps of grass. My backpack toppled off the seat and onto the floor as water bottles were jarred from cupholders. After a few seconds, we landed firmly back on the roadway.

"Weeeeeeeeeee," I announced. Kelby smiled and rolled her eyes.

We stopped for gas about twenty minutes later in whiteout conditions. I stepped out of the SUV wearing jeans and Sperrys, my glasses instantly fogging up in the frosty air. Amir and Youssef congregated to discuss options; apparently, Old Man Winter's off-season temper had closed the main road to the desert.

The pair decided to backtrack a half hour to take a different road. A hot jet of air blowing into my face from the heater lulled me to sleep; when I awoke hours later, we were in chaparral biome with short, stubby trees. The landscape was flat as far as the eye could see, enclosed by the dismal silhouette of distant mountains. Having woken up at 5:00 A.M. to write a *Washington Post* article, I was sleep-deprived to the point of incoherency, and it would be another eight hours before we'd arrive in Marzouga.

—∾—

The Earth is a rotating system. It's easy to forget that we live on a 7,918 mile-wide globe that turns around once every 23.93 hours. Everything on that sphere feels the Earth spinning, and is subject to the Coriolis force. That's integral to the general circulation of the atmosphere. (In the Northern Hemisphere, that Coriolis force pushes speeding objects to the right; whether you're a pilot crossing the Atlantic or a sniper aiming at a target a quarter mile away, you have to take the Coriolis force into consideration. Even home run baseballs deflect a fraction of an inch to the right.) The Coriolis force gets stronger as you increase in latitude.

Latitude also has a role in our climate. Every location on Earth receives roughly the same cumulative duration of sunlight every year, but not necessarily the same amount.

On the equator, every day is about twelve hours long, whereas some days at the north and south poles feature twenty-four hours of daylight or darkness. It mostly evens out in terms of timing.

What does vary is the angle and intensity of the light. The sun's most direct rays shine on the equator, which is why temperatures there are warm year-round. But at the poles, even the summer solstice is bone-chilling; that's because the incident sunlight irradiates at a low angle. As the Earth's surface curves away from the sun, the same beam of light is spread over a greater area, diminishing its intensity and warming effect. That imbalance drives the overarching air currents that drive what happens in the atmosphere.

Since warm air is less dense, the air over the equator rises. Chilly air near the poles sinks. That overturning means air from the equators has a tendency to ascend and move poleward before subsiding and returning to the equator. That's what's known as the Hadley cell.

Because of the Earth's rotation, however, air moving toward the poles curves to the right as the planet's radius shrinks. That boils down to the conservation of angular momentum.

Think of a spinning ice skater. If her arms are outstretched, she rotates slowly, but spins much more swiftly if she draws her arms inward. Poleward-moving air parcels, or pockets, at the upper levels of the atmosphere accelerate to the east as they migrate away from the equator. As their latitude increases, so too does their eastward velocity.

Theoretically, that would mean air at the poles should move infinitely fast. That can't happen. Instead, the regime of Hadley transport exists within about thirty degrees of the equator. Beyond that, air masses begin to pinch off and curl back on themselves, forming ebbs of high and low pressure that preferentially distribute heat to the poles. That's the area of eddy heat transport.

In the Northern Hemisphere, the Hadley cell's northern extreme is marked by sinking air. The same is true at the southern edge of the Southern Hemisphere's Hadley cell. Where air sinks, it warms and dries out. That's why northern and southern Africa, as well as parts of Asia; the Middle East; and most of Australia, Chile, and Argentina are deserts. The United States is in the right latitude belt to be mostly desert, but the Caribbean and Gulf of Mexico add enough moisture to the atmosphere to prevent that.

Near the equator, the rising branches of the Hadley cell converge into a broken band of showers and thunderstorms that encircles the Earth. Meteorologists call this annulus the intertropical convergence zone. It wobbles north and south each year as it follows the sun's richest heating.

Deserts are defined as areas that receive fewer than ten inches of average annual precipitation. Believe it or not, temperature is irrelevant in the classification—the world's largest desert is actually Antarctica. That's because the inhospitably cold air can't hold any moisture, so anything more than a brief light snowfall over most of the continent is rare. The same is true in the Arctic. Utqiagvik, Alaska, the northernmost town in the United States, situated 320 miles north of the Arctic Circle at the tip of Alaska's North Slope, routinely sees blizzard conditions, but the town only receives 5.4 inches of precipitation in a typical year.

The Sahara is the world's largest nonpolar desert, and its climate is closely tied to the sensitivities of the Hadley cell. Modern research suggests that the Sahara is expanding thanks to desertification driven by climate change. Some estimates state that it grew by 10 percent during the twentieth century.

Desertification is a self-reinforcing process. As temperatures warm, the atmosphere evaporates more moisture from the ground. That desiccates the landscape and kills vegetation on the fringes of the desert. With a drier ambient environment, the air can heat up even more. It's of grave concern in the Middle East and North Africa, commonly referred to as MENA, where water resources are already taxed and food security hangs in a precarious balance.

While in Rabat, we had spoken with an atmospheric research scientist who warned that farmers were struggling to keep up with the quickly evolving conditions. One apple farmer we visited had just finished installing a costly drip irrigation system to conserve water during the dry season.

A 2017 study in *Regional Environmental Change* notes that stream flow could drop 15–45 percent by the end of the twenty-first century, with extreme heat projected to perniciously affect a third of the MENA region. The paper estimates crop yields could decline by 30 percent or more, precipitating a mass exodus toward cities amid severe economic impacts. According to the authors, "a severe and sustained pressure on resources could contribute to further social unrest in the already unstable political environment that currently characterizes parts of the region."

~

We arrived to the desert long after the sun had set. As we descended out of the high terrain of the Atlas range, the sky abruptly cleared, the landscape transforming into a sandy, stony basin. Our drivers flicked on their high beams as they pulled off the paved road, speeding through the darkness. Amir and Youssef drove the SUVs like dune buggies; I could tell they were having fun. It wasn't their first rodeo.

At last, I spotted a pinprick of light in the distance; the only other source of luminance came from the stars above. It got brighter as we approached, finally revealing itself to be our accommodations for the evening: four luxury tents arranged in a semicircle around a campfire, each equipped with a comfortable mattress, electricity, pillows, blankets, and a rug. The door was just a flap of fabric.

How do they keep the bugs out? I wondered. Youssef seemed to know what we were thinking.

"There's not much that lives in the desert," he said. Our settlement was nestled in a small depression between fifty-foot knolls of sand. The air was cold and still.

I awoke early the next morning, shivering. I knew the desert got cold at night, but I hadn't been prepared for near-freezing temperatures. I lay in bed in a hooded sweatshirt and jeans. Steeling myself, I finally conjured up the verve to go for a walk.

The resplendent colors of sunrise were asserting themselves over the baronial blues of twilight. My sneakers sank several inches into the ground as I stepped off the burlap, canvas-like walkway. I bent down to examine the sand: it was powdery and dust-like, as fine as table salt. I began to understand how outbreaks of wind-lofted Saharan dust could occasionally make it as far west as the US Gulf Coast.

I climbed a wind-sculpted sand dune, a light, whispering breeze stirring loose granules within a few inches of the surface. I squinted.

There wasn't a structure, tree, bush, or vehicle to be seen. The vastness of the world's largest nonpolar desert—more than 3.5 million square miles—was paralyzing. I sat there pensively as the day began.

Temperatures spike within twenty minutes of sunrise, the frigid sand suddenly a bed of hot coals. I ditched my sweatshirt, opting for short sleeves and lightweight hiking pants. Even thick, dark sunglasses could barely handle the vigor of the sun's fulgent shine.

In the desert, you never stray far from your guide. It doesn't take long to become disoriented in the endless array of undulating hills, which can quickly prove fatal. Dehydration comes swiftly and without warning; the dry air evaporates sweat more quickly than it can accumulate, desiccating a person before they even notice they're sweating.

The eight of us saddled up for a camel ride immediately after breakfast, but I only lasted ten minutes—turns out that camel humps are a little bit more unforgiving on the male anatomy than, say, a lump of memory foam. It didn't help that my camel, whom I had nicknamed "Janet," was frothing at the mouth. Walking through the sand uphill was like climbing a mountain of molasses, but at least I wasn't wincing with every step.

Shortly after supper, I corralled Jason, Kelby, and Race, ushering them into our shared tent.

"Bedtime," I announced. They looked at me bewildered. It was only 7:30 P.M.

"In northern New Jersey, you can see a maximum of 50 to 100 stars," I explained. "In Oklahoma, maybe 500, and in Minneapolis, about 200 or 300." They nodded. "In the Sahara Desert, the lack of light pollution and moisture means we're close to the theoretical max: 5,400 stars. It'll be like you're on a different planet."

Begrudgingly, the trio slipped into their beds. I circled the room, inspecting each of the electrical sockets to see if they would accommodate a camera battery. Once I had plugged everything in, I crawled into bed. Sleep came quickly and without permission.

—w—

The clangorous din of my phone alarm burrowed its way into my dreams. Eleven o'clock had rolled around, and when I realized I had to be awake, I sprang out of bed. I had a rendezvous with the Milky Way, and I was not going to be late.

I flicked on the light. Jason groaned, yanking the covers over his head; Kelby sat upright, with frizzy, matted hair and a blank expression. She looked she should be driving the Magic School Bus. Race was nowhere to be seen.

Hastily, I collected my camera batteries, fastened a wide-angle lens onto my Nikon D3200 body, and grabbed my flashlight.

"Y'all aren't going to want to miss this," I said to Jason and Kelby. By now they were just barely stirring. "Bring a towel or something to sit on." I knew that three hours' worth of radiational cooling would make sitting on the sandy desert floor as harsh as curling up on an ice rink.

I stepped outside of the tent and glanced skyward. Only a few stars were visible, but that was because of the crackling bonfire. I sauntered over and stretched out my hands. Race was sitting next to Damien; Shelly and Karen were resting atop tree stumps opposite them. I wondered where the wood had been trucked in from.

"It looking like a good show?" Race asked. I nodded.

"Every night in the desert is a pretty good show. Maybe we'll even see a shooting star."

A rustling from our tent heralded Kelby and Jason as they stumbled into the darkness. Race stood up.

The three of us trudged up a sand dune, the fire's crimson glow disappearing as we dipped into the dune's lee. More stars emerged with every step away from our encampment. After the second and third dune, I felt like I was prancing through outer space.

"This is incredible," Kelby said, her mouth agape. A cloudy, nebulous blue arc bridged across the heavens, made up of individual phosphorescent grains. In it were a few scintillating kernels that shone brighter than the rest, twinkling as the light from centuries-old nuclear fusion reactions reached our eyes. We were staring at a cross-section of our 100,000 light-year-wide spiral galaxy. Jason and Race craned their necks skyward. No one said anything; the scene spoke for itself.

The sky looked like someone had spilled glitter onto a black backdrop. The stars seemed directly overhead, but just out of reach; if I got a running start, maybe I could catch one.

It required all of my concentration to steal my gaze away from the stars long enough to spread out a towel I had snatched from our camp. I unfolded it on the cold, hard ground, and sat. The others joined me. It wasn't large enough for us all to cram onto, so we alternated direction and lay with our backs, shoulders, and head on the towel; our outstretched legs dangled in the sand.

"Ooh, ooh ooh!" I shouted, suddenly unable to speak. I pointed at the sky, but it was impossible to miss what had captured my attention: a ball of fire was igniting the zenith, streaking overhead before fragmenting into a spattering of flaming red, orange, and yellow chunks of debris. I had never seen anything like it; Kelby gasped, our vociferous shouts echoed by distant cries presumably from the others seated around the campfire, two hundred yards away.

"What *was* that!?" Jason asked, shouting in awe. I pondered for a moment, still thinking through what I had seen.

"Either it was a bolide, or a large exploding meteor," I said, "which would be pretty rare and special. We don't have any way of knowing if it hit the ground. Basketball-sized meteors strike Earth about once a month. Otherwise it may have been a satellite or piece of space junk breaking apart in our atmosphere. The frictional drag from air resistance would be forceful enough to cause it to explode and burn up."

"Have you ever seen anything like that before?" he asked.

"Honestly, no," I said. "But I'm glad you guys were here for it."

We all went back to staring skyward. Every minute featured something new: a faint, blinking satellite zipping across the sky; gleeful bellows when we spotted a new constellation; or simultaneous "oohs" and "aaahs" whenever we witnessed a shooting star. Even though there weren't any major meteor showers going on, I had never seen so many.

Meteor showers occur when the Earth passes through a stream of debris left behind by a comet, asteroid, or other celestial body during our annual orbit about the sun. (That's why meteor showers, like the August Perseids or the December Geminids, happen at the same times every year.) The interstellar pebbles, often only the size of a grain of puffed rice, burn up in Earth's outer atmosphere, combusting as they encounter air while moving at speeds of forty miles or more per second. Imagine driving through a swarm of bugs; when you hit one, it leaves smear across your windshield.

The night was quiet, a blissful solace enveloping us. I was surrounded by nothingness for hundreds of miles in all directions, staring up at specks of light orbiting through a universe of nothingness. It felt an odd contrast to be a small island of consciousness, deriving significance and meaning from our collective somethingness.

In the grand scheme of things, I was hardly any different from the grains of sand resting beneath me. Yet the feeling of insignificance was comforting. Moments like that are a reminder of the brevity of life, and how small of a role we play. I know my time here is fugacious; that's why we have to make it count.

The final month of study abroad took us to Bolivia, where Heather was expeditiously fired as soon as the plane landed. (I may or may not have had something to do with it, though the fact that she abandoned us in the middle of Paris at 1:00 A.M. without cell service during a botched airport transfer didn't help her case.) While that didn't prevent further asininity on the program—like when we held a ritual sacrifice of leaves around a campfire while singing about llamas "to appease Mother Earth"—things were at least getting better.

We spent two weeks in La Paz, the highest administrative capital city in the world. The runway at nearby El Alto International Airport is 13,235 feet above sea level. Colorful skyscrapers clogged steep mountainsides, an elaborate gondola network the city's main vessel of public transportation. The streets were too narrow and inclined for buses, and a subway could never operate.

At two and a half miles above sea level, air pressure had dropped by a third. The thin atmosphere meant that short uphill walks were laborious and exhausting, and two glasses of chardonnay were enough to put anyone to sleep. Some of my peers chewed on coca leaves, rumored to help with altitude sickness.

Though the program had gone off the rails, we did have one source of knowledge and a confidante. His name was Alan Steinwick. The program had hired him to teach environmental sciences classes, and while I didn't learn anything new scientifically, Alan had a knack for knowing how culture and science intertwined.

We took a six-hour journey to the Isla del Sol, or Island of the Sun, a remote island in Lake Titicaca. Getting there required a lengthy boat ride, and there were no vehicles, stores, or internet. Instead, we occupied small dwellings with local families, each of whom owned multiple adorable alpacas. Alan brought us to the sites of Inca ruins, explaining how ancient civilizations were able to draw inferences about the weather and climate. The more I learned, the more my skepticism vanished.

As it turned out, the millennia-old Inca practice of predicting rainfall through taking observations of the Pleiades, a star constellation, had merit to it. Modern scientists have since been able to tie cirrus cloud cover, which affects whether the stars are visible, to the status of El Niño-La Niña, which is known to influence rainfall over the Andean Plateau.

Though our technology has come a long way since then, there's still a lot we don't know. Indigenous practices for managing and preserving native flora, fauna, and maintaining healthy local ecosystems are gaining long overdue acknowledgement worldwide. Once in a while, it might be helpful to look backward to guide our path forward.

Graduation

After studying abroad, the calendar flipped to 2019, and with the New Year came my final semester at Harvard. It was hard to believe the finish line was in sight after a four-year-long uphill battle to achieve an atmospheric sciences education. I was ready to be done with differential equations, eigenvectors and the Ertel potential vorticity equation, and, for the first time in my life, I was doubting my dreams.

My friend Misha had landed a six-figure job leading a team of cancer researchers in Boston. Camillo would be heading to pursue a physics PhD in Canada, and Tess was returning to Harvard for graduate school. Most of my other friends were traveling the world, gearing up for prestigious consulting positions, or moving to Silicon Valley. Kelby, after returning from Bolivia, was applying to law school.

I, on the other hand, hadn't been recruited; in fact, no one would even return my calls. One television station in a low-tier Wisconsin market rejected me within four minutes of receiving my application, even though my demo reel was eight minutes long. Nobody would take me seriously.

My wine-scented CamelBak water bottle had made its way back from study abroad; I found myself refilling it more frequently to get through the days. *If you keep relying on this as a solution, it's going to become more of a problem*, I thought. But it was the only thing that offered a reprieve.

A Dixie cup of chardonnay before brushing my teeth at night became a ritual lullaby of temptation; it was a way to stave off my inadequacy and finally fall asleep.

Conversations at the lunch table that once centered on schoolwork were replaced by students comparing offer letters or contracts. Depression set in by March. I flew to China during spring break to deliver motivational speeches (a job I'd picked up junior year), mainly for the confidence boost, but it didn't work; I had nothing to show, no cards to play; my hand was empty.

After returning from China, I took a last-minute flight to Fort Myers, Florida, to visit my grandfather. He had been my hero since birth; my parents and I lived with him until I was three. Even at eighty-six years old and hardly able to speak, he still made me laugh; we could have a conversation without words.

I woke up at 3:30 A.M. to board a Spirit Airlines flight from Boston to Southwest Florida Regional Airport. Tuesdays were my only day without classes or work, so I decided to squeeze in a day trip in between. It was four hours each way; I crammed textbooks, atmospheric chemistry problem sets, and my laptop into a tattered backpack and took to the skies.

I landed at 9:00 A.M. and took an Uber to his mobile home; it was unlocked. My grandmother had passed away when I was seven, and Nancy, his girlfriend of more than a decade, had been with him ever since. She knew I would be flying down; he did not. She was out golfing with her friends Mary Ford and Chuckie.

I strolled into the cramped kitchen, moved a pile of newspapers to one side of the dining table, and plopped my laptop down. Then I brushed aside a litter of stray candy wrappers (Nancy loved hard candy) and got to work. The cluttered room was silent, save for the metronomic ticking of a hummingbird-themed clock mounted between two wooden and Plexiglas china cabinets; the wallpaper behind it was a faded yellow.

Around 11:00 A.M., Papa made his appearance, stumbling out of his bedroom groggily like a bear awakening from hibernation. He looked frail and weak, but his eyes still twinkled with wisdom and mischief. He turned and noticed me sitting at the table, appearing confused before a slender smile crept across his face.

"What the hell are you doing here?" he asked in disbelief. He looked like a child who had been visited by the tooth fairy.

"I was in the neighborhood, so I figured I'd swing by," I said with a laugh. "Whaddya say we go grab some lunch?"

He shakily withdrew to his bedroom to get ready and returned in a Hawaiian shirt and blue jeans. I had never seen him wear anything remotely colorful—it was usually all plaids or white tank tops and suspenders. It was like he was celebrating.

I guided him out to his minivan, which was filled with errant golf clubs, Altoid tins, Glen Campbell CDs, and fast food receipts. A purple rabbit-ear keychain dangled from the rearview mirror. After buckling him in and making sure he was secure, I walked back around the pockmarked van. A new dent caught my attention.

"Nancy backed into some wall thing at the bank," he said as if reading my thoughts. I smirked at him.

"Sounds like she took 'drive-through bank' a bit too literally," I joked. He laughed.

Nancy, four years younger than him, was a professional grandmother. She took her job seriously. She had four grown grandchildren and a dozen great-grandchildren. Every soccer game, birthday, and school play was adroitly plotted on her calendar. She kept tabs on my grandfather with an equally watchful and authoritative eye.

Nancy first entered the equation when I was nine. She had baked (or, in reality, cremated) a tray of preformed chocolate chip cookies. Striving to be a polite fourth grader, I complimented her on them. That locked me into an inescapable pattern of receiving a charred heap of chocolate-chip charcoal nuggets every time I saw her.

On the short drive to Jack's Diner, I realized it was the last time I'd see my grandfather; he wouldn't make it home to Massachusetts that summer. In retrospect, I think he was aware, too. We reminisced about my youth, him recounting stories of me in the backyard strawberry patch.

The 1970s-style diner was largely empty; a jukebox stood adjacent to the retro ice cream counter. I helped my grandfather into a booth, easing him onto the maroon-upholstered seat. The slim, curly haired waitress, probably in her early sixties (which made her practically a stripling in Fort Myers, where the median age was pushing eighty), winked at me, and smiled as she walked over to take our beverage order.

"What'll it be, gents?" she asked. I smiled patiently at my grandfather.

"Nancy said I can't have orange juice," he grumbled, reverting to the six-year-old caricature that sometimes showed itself when his blood sugar dropped; it was symptom of his many strokes. "She says it's too much sugar."

I leaned toward him and slapped the table while grinning, turning to the waitress.

"We'll take the biggest, sweetest orange juice you have," I said. "Two of them, please." He grinned.

The afternoon flew by. He didn't eat anything at Jacks—he couldn't swallow much by then—but I could tell it was the most nourishing meal he had had in months. Afterward, we returned to his house, strolling around to the garden. His favorite pastime was narrating the handful of overgrown plants that the small patch of dirt contained. I knew I would miss this more than anything.

"Excited for graduation?" he asked as we sat quietly in the shade. The midafternoon sun was bright, but not hot: the temperature seemed just right. Everything was just right.

"Not really," I confessed. I could tell he was searching for his words.

"It's all part of the ball game," he replied. I stared at the ground and smiled somberly.

"You coming to graduation?" I asked. Harvard was the only college whose name he knew.

"I don't know about that," he said; I knew exactly what he meant.

"Well if you're not here for it, I still want my gift," I said, smiling. My eyes were watery, but my sunglasses hid them. He turned to me, rolling his eyes and opening his mouth, feigning shock and surprise.

"Whaddya want?" he asked. "A vanilla cream doughnut?" I was surprised he remembered—it was a staple of my childhood.

"Actually, a tornado," I joked, trying to keep my voice from cracking. "If you have that power by then."

"How about *two* tornadoes," he said hoarsely, coughing in between shallow breaths and laughs.

"Deal," I said. I gave him a hug, grabbed my backpack, took an Uber to the airport, and flew back home. He was gone three weeks later.

The Ultimate Gift

April rolled around and I was grasping at straws, frantically reevaluating my future. I decided to consider a last resort I had adamantly sworn off: going to Wall Street. I would be a guppy in a sea of sharks.

I emailed an alum from the class of 1986 who I had become friends with during my junior year. We had met at Harvard's Celebration of Scholarships dinner, where I was a keynote speaker. It was a suit-and-tie catered dinner; in attendance were four hundred of Harvard's most generous donors, representing a net worth of tens of billions of dollars. This man, named Patrick, had launched his own investment banking firm, owned a dozen houses, and had $50 million in the bank. Most importantly, however, he had connections.

I knew that many hedge funds had a need for meteorologists who could offer insight into how upcoming weather trends would influence consumer purchasing patterns or demand for commodities. It was a highly profitable industry. Patrick linked me with the heads of several of the nation's largest hedge funds; I dusted off my suit and tie and, and the end of April, sauntered into Citadel in Boston's Financial District for an interview.

It turned out to be a Zoom interview. I was herded into a small conference room and told to wait. Minutes passed before the wall-mounted monitor flickered to life. A man sitting at a desk appeared on camera.

Without any pleasantries, greeting, or exchange of information, he began parsing my resume.

It was a whirlwind. I left the office thirty minutes later, processing what had just happened. I hadn't been provided any information about the role to which I was applying; I was told that "Citadel doesn't advertise roles. We create roles for people we want." Things were going well until I was asked about my coding experience; the interview ended abruptly after that.

By May, I was desperate; I wanted to look forward to my forthcoming storm-chasing adventure, but the surmounting uncertainty of my future was a haunting distraction. The final five days of school were set to be jam-packed: I had three problem sets and two final projects to do as well as a term paper for a course at MIT that I hadn't even started; I also was set to deliver a keynote address for a scholarship luncheon on Sunday, May 12. Then I received a last-minute invitation to a job interview at D. E. Shaw in New York City on Tuesday, May 14.

This seems promising, I thought. Surely a major hedge fund wouldn't be bringing me to the Big Apple unless they were interested, right? I snagged a predawn ticket on the Amtrak Acela, ironed my suit, and scooped up my bundle of loose papers and binders. I took an Uber to South Station in Boston at the crack of dawn, bought a doughnut and, after four hours, floated into a pristine Manhattan skyscraper. The building's lobby reeked of perfume and opulence; the street outside of pizza and stale urine.

Five interviews and a bowl of soup later, I was feeling pretty confident; we were talking target salary, base cities, and available start dates. I had "everything [they were] looking for." Plus, I was in a good mood anyway; model data I reviewed on my lunch intermission was highlighting a potential days-long severe weather outbreak on the Great Plains at the end of the week, and tornadoes appeared virtually certain. One of the interviewers, a man in a striped button-down shirt in his mid-forties, asked me about storm chasing as he walked me out to the elevator late in the afternoon. He was wearing a Rolex.

"So, I've got to ask," he broached casually, half smiling. "You're passion is clearly the weather. Are you going to be as passionate about this?"

I laughed, pressing the button to call the elevator.

"There's nothing that rivals how much I love weather, but any time I set my mind to something, I do a damn good job."

He nodded and smiled, seemingly satisfied with my answer. The elevator dinged.

As he walked away, he paused for a moment, then spoke in a hushed tone.

"Just don't give up on doing what you love. Trust me."

The sleek, metal doors of the cold elevator slid together smoothly.

Wednesday, May 15, was my last day of classes at MIT. It was reading period at Harvard, which means no classes or assignments the week before finals. My train back from New York didn't arrive until close to midnight, and I had been up until 4:00 A.M. working on a final packet of atmospheric dynamics questions for a class at MIT. I still hadn't even begun my four-thousand-word paper on microscale wind effects within tropical cyclones. It was due at 11:59 P.M.

My dorm room was bare: every weather map, poster, and glass barometer had been packed into crates and driven home to Cape Cod. Patches of paint were missing from the walls where my Command Strips had been adhered. The desk, bureau, and closet were empty. I felt the same.

I spent the morning hours at MIT, frantically typing my final paper on tropical meteorology while making appearances in my last two classes of undergrad. By midafternoon, I returned to Currier House at Harvard. I silently collected my lone suitcase, slid my key under the resident dean's door in an envelope, and exited the building to stand by the curb for the very last time. My Uber pulled up moments later.

I hastily threw my suitcase in the trunk of the sleek blue sedan, but something caught my eye: a soothing brush of gentle pink painted on a tree next to me. I looked up and realized all the cherry trees that flanked the brick and cobblestone driveway were budding; some were opening their blooms, the mild, comforting air tinged with sweet scents.

I paused for a moment, surveying my surroundings; the chatter of birds was accompanied by sounds from the engine of a receding Quad to Harvard Yard shuttle bus. A gentle breeze stirred the coral-colored trees, producing a shower of matrimonial petals. Without the constant stress of classes and homework, it was the first time that I had paused to appreciate the beauty that had surrounded me for four years.

Slipping into the Uber, I thought how quickly the time had gone, questioning if I had made the most of my time at Harvard, and wondering what was next. The bustling cluster of dormitories grew smaller as the vehicle unceremoniously crawled down Linnaean Street. I had my boarding pass queued up on my phone, but, in reality, I had no idea where I was going.

"Want me to ask if we can sit together?" I asked Michael. I found him at an airport lounge in his signature maroon hooded sweatshirt and wrinkled khakis. A pair of luxury noise-cancelling headphones wrapped around his neck. He was watching Showtime's *Weeds* on his Dell XPS15.

"We'll be on the same plane anyway. I'll see you in Oklahoma," he said, shrugging. It was a valid point. Besides, I had another 2,300 words to write in four hours' time.

After a connection in Atlanta, we hopped aboard a Delta "Mad Dog," or McDonnell-Douglas 88 aircraft. It was a sturdy, reliable plane known for its loud, screaming engines located on the tail of the fuselage. It would take us to Oklahoma City.

The final keystrokes and taps on a sweaty mousepad at 10:52 P.M. Central Time signified the end of my formal education. We had just flown over the Mississippi River when I paid for an hour of in-flight Wi-Fi to submit my essay online. I composed an email to Kerry Emanuel, a famed professor of atmospheric sciences at MIT, attached my fourteen-page write-up, and clicked SEND. *Well that was anticlimactic*, I thought.

We landed at Will Rogers World Airport around midnight and took an airport van to pick up my truck; I had driven it down from Massachusetts two and a half weeks prior. An hour later, I was trudging up a flight of stairs at the Super8 Motel in Moore, Oklahoma, with Michael in tow. He was carrying a Tupperware of congealed gummy worms.

"This is neither super nor an eight," I muttered, hoping the universe would hear my attempt at comic relief. "It's like a mediocre six."

—

At 4:00 A.M., I woke myself up to check the weather models on my iPad. I glanced over at the other side of the room. The orange fluorescence thrown by a streetlight was peeking through the window blinds, faintly illuminating Michael, who had rolled himself in blankets like a burrito. I knew he'd probably want me to take him to Taco Bell for lunch.

By 8:00 A.M., I was up. The next day, Friday, was shaping up to be a big day, and I had no time to waste. While Michael was still dozing, I was using Google Maps to virtually drive the main roadways in southwest Nebraska and northern Kansas. I had a feeling that would be my target area. I strolled out to Ridgie to retrieve a pair of state atlases before returning and draping the crinkled paper maps across my bed.

I settled on Kearney, Nebraska, as a place to stay that Thursday night. It was a seven-hour drive from Oklahoma City, but the fact that we had one cushion day before the outbreak of active weather commenced was ideal. I booked the Quality Inn and Suites for $71, flipping up my email to double check the confirmation. A message from the New York hedge fund stood boldly at the top of my inbox. I clicked on it, a frenzied trepidation coursing through my veins.

"We regret to inform you . . ." the two-sentence email began. I stared at my laptop's dusty screen in disbelief, my breathing becoming conscious and labored. My last shot at a job had just evaporated. *But they loved me. No one makes it to all five rounds and then gets rejected.*

I silently closed my laptop and grabbed my room key before slipping outside into the bright midmorning sun. My eyes stung, the sun's brilliance slicing through the salty tears like daggers. I was staring into a warm, blazing kaleidoscope.

~

"There's the train!" I hollered, briefly forgetting the climaxing sense of failure that had been consuming me for months. Maybe I was numb to it by now, or perhaps my internal eight-year-old was fixated on the screeching freight train clattering across the street in front of me as a modicum of relief. The alternating red flashers and twinkling lights on the gates were indeed a hypnotic, and welcome, distraction.

"You really are a child," Michael said.

"Oh come on, you don't like trains?" I prodded jokingly.

"No. They're big and loud and make me wait when they cross the road."

We had been driving for seven hours, powering on to Nebraska nonstop after a morning meal at Waffle House. Michael described the food as "cheap, greasy, and American." It was tough not to take it personally.

The air was cool and dry as we parked and walked next door to the Whiskey Creek Wood Fire Grill for supper before returning to the hotel. Michael fell fast asleep while I untangled a web of wires, charging a dozen or so cameras before finally shutting my eyes. I would awaken four times overnight to review incoming data.

"I'm thinking McCook," I said to Michael, who nodded approvingly. It was late morning and the chips were down: a big storm system was brewing. Low pressure was sliding out of the Rockies, strengthening southerly flow. That brought surging temperatures and humidity. I packed up the truck and made sure I had a pair of safety glasses for Michael. Then we headed off in search of a quick lunch.

"Look, they have a train museum!" I said eagerly, a yellow-and-red Union Pacific railcar stealing my attention as we drove slowly down West 11th Street. "We have some time to kill. Want to go?"

"You can go. I'll wait in the car," Michael replied matter-of-factly. I frowned, but decided not to press the issue. We drove to Taco Bell instead. He ran inside for food while I began unhooking the stowed hail cage that hung out of the tailgate.

The hour and forty-five minute ride to McCook was a race against the clock. Storms were set to fire any minute; high-altitude temperatures ahead of the approaching low were cooling fast, and the contrast against the warming surface meant pockets of air would ascend rapidly.

Ground-level observations told me there was a weak boundary, or subtle change of temperature and wind direction, draped somewhere near McCook in southwest Nebraska; that would enhance low-level convergence, the gathering of air, and help trigger the first storm.

My forecast panned out exactly as I had hoped. As we pulled into town, the western horizon was already beginning to darken. Glancing at radar data on my phone, I decided to position us four miles west of McCook in the town of Perry. Michael was rapt with attention, his eyes trained on the sky.

"Are we going to get big hail?" he asked. I surveyed the last radar scan I had for about a minute before answering him. Cell service had dropped out, meaning I'd have to make my judgment calls visually.

"I'll tell you what," I said, focused on a growing splotch of purple that had been acquiring a kidney-bean shape. "If we stay here, we'll get large hail to about baseball size. We can do that, but if we get a tornado warning, then we've got to move. I need a visual on the storm."

We were positioned where the storm's downdraft of rain and hail were going to sweep through, meaning visibility would drop to near zero in blinding downpours. If the rotation tightened and a tornado formed, we might not see it until it was too late. For now, only a severe thunderstorm warning was in effect.

"EEEEEEEEEEEK!" my weather radio screamed three times—it was my backup data source. An emergency alert was being issued.

"The National Weather Service in Goodland has issued a tornado warning for west central Red Willow County . . . and southeastern Hitchcock County," the radio announced. McCook was in the center of the warning. There was my answer.

"Are we going east?" Michael asked.

"Yep!" I replied, shifting into drive.

A gloom had descended on the midafternoon landscape. It was just after 5:15 P.M., and the towering thunderheads to our southwest were throwing shade—literally. I reached above the steering wheel to turn on the headlights. A light rain fell as we clipped the eastern fringe of the storm.

"There's the wall cloud!" I said to Michael. I turned off the paved roadway onto a dirt driveway and parked. Frayed strips of cloud cover swirled into the turbid base of the storm, drawn toward a shaggy appendage that hung just above the ground. It was getting close.

After five minutes or so, I got the nagging feeling that it was time to go.

"We've got to bail east," I stated. The wall cloud was virtually on top of us. We raced east two miles on the dusty country road network before following the grid north.

"Look at the clear slot!" I exclaimed, pulling over. A coal-colored cloud stretched from directly overhead to about two miles to the west, but beyond the edge of the massive self of mist hung broken mid-level clouds and daylight. A wedge-like void of dry air and white cloud cover was punching into the storm's rotating updraft, creating a collar of luminance as it tightened the tilted axis of rotation. It was the final step necessary for tornadogenesis.

"There's the funnel!" I said to Michael, who stood with his iPhone facing directly at the feature. The narrow gray tube was only a couple hundred feet above the ground.

"Let's go!" I shouted. Within seconds, we were racing down the road as my engine groaned, paying careful attention to stay within a few miles per hour of the speed limit. *Don't get tunnel vision*, I reminded myself. We encountered one traffic light, waiting for a fifteen-second millennium, and turned north.

"LOOK AT THAT!" I screeched, handing Michael my camcorder to mount to the dashboard. An elegant stovepipe tornado was skimming along the ground like the snout of an anteater, kicking up dust about four miles to our northwest. We were approaching from the south.

Michael, entranced, sat staring at the tornado.

"Here, record this," I said; Michael had agreed to film whatever we encountered as a courtesy since he was getting a mostly free trip. I was focused on the road. I turned to ensure he was affixing the camera to the dashboard tripod mount, but instead he was haphazardly holding it with his left hand while staring at his phone in his right.

"No snaps," I stated, noticing that my $1,000 camera was being shakily pointed at the rearview mirror and the overhead visor.

"I am holding the camera well," Michael barked. "I am taking pictures. It's not the same as sending to my friends."

This is going to be cause for a lecture later, I thought, though I was more focused on the tornado; it was now tilting and elongating, a sign that its ground circulation was lagging behind its connection to the cloud above. That was stretching the funnel, ordinarily the first indicator of weakening. I pulled over and gestured for Michael to follow me into an adjacent field. I was already standing across the street by the time he unfastened his seatbelt.

"Alrighty guys, big tornado!" I bellowed to the camera, cognizant of the fact that the video would be ideal for my demo reel—if I ever made it into television, of course.

The funnel was roping out as a surge of cool air acted to wither its circulation, allowing me just thirty seconds to snap photos before the *Wizard of Oz*–like twister began retreating into the clouds.

"Back to the truck!" I shouted to Michael.

"Are we in a hurry?" he asked.

"This could be the start of a tornado family," I said, flinging my wide-angle camera onto his lap. Radar indicated a second storm had fired on the first cell's flanking line, or a strip of clouds in varying stages of growth that forms like a tail attached to a thunderstorm. Northeast was the best option.

Fifteen minutes later, we were in a storm chase Bermuda Triangle, navigating on a narrow road through the rolling Sand Hills of Nebraska having lost internet connectivity once again. Even my radio didn't work. I suppressed my growing nerves to maintain a stern but controlled facade for Michael.

"Keep an eye out the right window," I said. "If the rain starts to lift and you see anything, let me know."

Except the rain didn't lift: it grew more intense and began to mix with hail. Within seconds, chicken egg–sized hail was lapidating the truck, but it was falling from the left (north). That wasn't a good sign. We were entering the core of the storm from behind.

"We must be in the rear flank downdraft, which means we are super close to the circulation," I said to Michael. Static blared over my radio as I silently pleaded for a signal. We were in between severe thunderstorms in an environment conducive to quick-forming tornadoes and we had no way to receive any data. A preloaded offline Apple Maps capture testified that the only available road options were forward or backward. Onward it was.

It was hardly 6:00 P.M., but night had fallen beneath the rotating supercell. I knew that the day's setup didn't favor rain-wrapped funnels, so my anxiety about driving into a tornado was relatively low, but being close to a tornado without the ability to see it is jarring. It's like sitting in a pitch-black room and knowing a tarantula is on the loose somewhere nearby. Suddenly, a garbled voice broke through the cacophonous rasps of the radio.

"A confirmed tornado was located near Stockville, or ten miles east of Curtis, moving northeast at 25 mph," urged the voice.

"Where are we?" Michael asked warily.

"Stockville," I replied softly, eyeing the GPS. He shifted uncomfortably in his seat, visibly nervous. I was too, even if I didn't show it.

The dashboard thermometer displayed an outside temperature of fifty-eight degrees. Gripping the steering wheel with my right hand, I pressed my left index finger against the glass; yep, fifty-eight was about right. It was loud and cold inside the vehicle, the air conditioning on full blast to prevent condensate from forming on the windows.

"It almost looks like there is dirt over there," Michael remarked, signaling for me to look out the right window. I turned.

"Michael, that's the tornado!" I said, both frustrated and relieved. We were two miles due north of it; I felt safe.

I quickly located a muddy turnoff that lead to a small parking lot next to an agricultural outbuilding on a hill. This would be perfect. I shifted into park and reached for my camera gear but heard a thump on my window.

"I'm stuck!" a man in a windbreaker yelled. He was wearing glasses and sandals and was standing next to my door. "Can you help pull me out?" Presumably a new storm chaser, he was driving a Toyota RAV4 that had fallen victim to the sloppy, sodden ground.

"I'm sorry," I replied, surprised the man wasn't waiting until after the tornado to take care of this. "I'm not mechanically inclined."

Without waiting for a response, I grabbed my camera and sprinted toward the neighboring field. Lime-sized hail was hammering my hard hat; I knew I'd end up with a few bruises on my back. I framed up a shot of the tornado whirling through fields to our south before running back to the truck.

"Come on!" I yelled to Michael, shouting for him to grab the camera. He was wearing flip-flops and clearly didn't want to venture into the sheets of rain. I couldn't care less.

"Let's go!" I emphasized. "The camcorder is already recording and zoomed correctly. Just take a couple steps back."

"I can't—," he began, sidestepping a mud puddle, but I was already mid-sentence as I launched into my explanation for the camera. As soon as the video was over, he retreated to the vehicle like an overtired dog trying to escape the rain. I couldn't, and wouldn't, move.

The tornado began to contract in diameter, its condensation funnel becoming wobbly. I closed my eyes and breathed in. Amid the chaos, I smiled at the sky and laughed solemnly, patting a nearby fencepost delicately.

Two tornadoes. I had graduated. Papa, somehow, had delivered.

Waffle Home

When you're in tornado country, Michelin star restaurants aren't really a thing. Subway, Sonic, Fuzzy's Taco Shop, and McDonalds are usually the only options in town. In smaller towns, especially across western Kansas and Nebraska, the only food to be found comes from gas stations.

Despite what you'd might expect, I love it. The five-year-old in me still gets giddy getting to eat junk food. And, on special occasions after a big chase, I get to eat at my all-time favorite restaurant: Waffle House.

They're everywhere in the South. Like an oasis of nourishment in an otherwise empty landscape, they pepper the countryside, open twenty-four hours a day to serve up cheap, greasy food made with love . . . and butter. There's nothing more comforting than a bacon grilled cheese sandwich and a double hash brown. It's a judgement-free zone in all regards. I want Waffle House to cater my wedding.

I've rolled up to a Waffle House soaked in rain, covered in leaves, and with a car door sagging off its busted hinges. I've trudged in there at two in the morning after a forty-hour marathon stint covering a hurricane. It's my go-to restaurant for good luck before a chase and for a celebratory meal after the storms. And even though I always get the exact same thing, I've somehow never paid the same amount twice. Besides, there's no greater confidence boost in life than being called "sugar" by a southern waitress in a rural diner.

Simplicity is the name of the game during storm chasing, including at hotels. Instead of Hiltons and Marriott, motor lodges and motels are usually the best options in town. As long as I can smuggle a couple extra mini soaps and shampoo with me as souvenirs, I'm good.

Naturally, some hotels are nicer than others. I once stayed at one with shattered windows and a wall that I'm pretty sure was bloodstained; it only cost $29 a night, however, so I couldn't pass it up. A storm rolled through late that night, knocking out power to the detached building in which I resided. I couldn't shake the feeling that I was being watched, lightning flashes through the windows casting odd, person-shaped shadows on the wall.

Others are cheap, too, their prices not reduced due to double occupancy with the supernatural, but just because they're run-down or vacant. That's the case in towns like Shamrock, Texas, once a booming town along Route 66. Now the community is largely a ghost town, with more than three hundred hotel rooms for only a small handful of weary travelers.

It's easy to score a $25 hotel room there. (Last time I did, someone had snaked the refill line to the toilet outside the tank, so every flush was met with a geyser of water across the room.) The town is also home to Hasty's Diner, a family run watering hole that serves four types of French fries. (I'd be lying if I said I hadn't feasted on all four in one sitting.)

Across the street from Hasty's sits a combination Chevron/Taco Bell, a gathering point for storm chasers on the afternoons of chase days. Its position next to north, south, east, and west road options offers the perfect launching point at the start of a chase. Chasers can fill up their cars before storms fire, while they themselves depend on tacos for fuel.

On big days, the dirt parking lot behind the drive-through turns into a veritable chaser tailgate, with dozens of gangly, shaggy-haired men and a handful of women in baseball caps holding cell phones and squinting at the sky. A lone cloud will appear, prompting some of the chasers to excitedly congregate while others glance even more intently at their phone screen.

One by one, doors will slam as vehicles exit the parking lot. Every chaser pays attention to which direction those departing head: *looks like Shaw's playing the triple point to the north* or *Timmer's aiming for tail-end Charlie*. Eventually, I take off, too. I try not to look in the rearview mirror, crossing my fingers that I've made the right choice.

Up Above the World So High

I'm getting nervous. Are you nervous?" I asked. It was 4:10 P.M., and I was becoming increasingly concerned. We still had another twenty minutes to go, but the sun was just above the mountains.

"I'm worried, too, but I think we're good," Dan said, hesitating while eyeing the distant hilltops. "Yeah, we're definitely good." We exchanged an anxious glance, sporting raised eyebrows and wide smiles.

I took a deep breath, trying to calm my nerves. The dry, clean air was thin, but it felt good. It was July 2, 2019, and I was perched atop a mountain on the western edge of the Chilean Andes. With me was my good friend and colleague Dan Satterfield, a veteran meteorologist and my solar eclipse buddy. Dan had turned sixty just two days before, but cascading flight delays left him celebrating with a stale pack of Oreos on a rural runway in Mendoza, Argentina. Happy birthday indeed.

I, meanwhile, had graduated college barely a month prior and had just returned from my Great Plains storm chase with Papa's miracle double tornadoes. And I had my new job at the *Washington Post* lined up. It was a prestigious company, but I wasn't remotely excited.

I wasn't going into "the biz"—my only TV offers had lower salaries than I'd made waiting tables; instead, I was moving four hundred miles to work in an office building. Maybe I wasn't good enough. I felt like abject failure.

But I didn't have time to dwell on it. Dan and I were just over fifteen minutes out from the moment we were waiting for: totality. The late afternoon shadows were becoming sharper, and the landscape had an odd sepia-like hue to it. Cacti stood sentry on the hillside we had summited, the dusty valley below studded with scrub brush and boulders.

It would be Dan's and my second solar eclipse, coming 681 days after our first encounter with the moon's shadow. Dubbed the Great American Eclipse, the August 21, 2017, occasion had been marked by totality tracing a seventy-mile-wide slice across the Lower 48. Millions migrated into the path to gaze skyward, unaware that their lives would be changed by what they saw.

Shivers ran down my spine and my hands shook that day in a rural Nebraska field as I watched the sky transform into something otherworldly. After totality ended, I recall turning to Dan and joked, "Ready for July 2, 2019?" We both knew we could never miss a moment like that again.

Nearly two years later, I found myself looking up at the sky again and smiling; not a cloud could be seen. Now that we were finally here, we could breathe a sigh of relief. Getting here had been a marathon, but it was about to be worth it.

I had reserved our hotels in September 2017, and booked flights in late 2018. In the months leading up to the trip, I spent countless hours poring over meteorological, astronomical, and topographical data, plotting our every move. When the time came to actually board my flight to South America, I couldn't believe that, after more than a year of planning, I was about to fly ten thousand miles round-trip to witness a moment two minutes and nine seconds long.

Arriving in Santiago four days before the eclipse, I was amazed. The ordinarily tranquil airport was bursting at the seams with eclipse chasers and astrotourists. I heard at least seven or eight languages, and measured the customs and declaration line at more than a half mile long. The air was buzzing with excitement and a sense of wonder, the word *eclipse* echoing from everywhere. I checked my phone in anticipation of rendezvousing with Dan, who was set to arrive within thirty minutes of me.

But American Airlines had different plans. Forty-eight hours of flight delays later, a weary but nevertheless undiscouraged Dan finally made it. I had spent the prior day and a half, including thirty-one hours at the airport, frantically making rearrangements, scooping up one of the last rental cars, and navigating Santiago's bus system like a contestant on *The Amazing Race*. (I even sent my former Spanish teacher a thank-you note.) After resting for four hours, we piled into our red Ford Explorer for the 290-mile drive to La Serena.

The journey north was serene. En route, we passed hundreds of people crowding the edge of the highway in small groups, touting backpacks and cardboard signs that read ECLIPSE. The city of La Serena, ordinarily home to two hundred thousand residents, was expecting half a million visitors. Traffic jams clogged the roads heading into town, street vendors weaving between idling vehicles to peddle eclipse glasses.

Attempting to catch some shut-eye that night was like a child trying to sleep on Christmas Eve. I was exhausted, but my mind was racing. I had vicariously driven every possible route on Google Maps a month earlier, accounting for sun angle, cloud cover, coastal fog, and just about every other variable imaginable. I slept from roughly 2:00 A.M. to 6:00 A.M., but as soon as the sun was up, I was, too. I knew that just hours later, that sun would vanish before our eyes.

We left our Airbnb at 7:00 A.M., descending twenty-one floors down to the lobby of our rented accommodations. Though La Serena is on the ocean, it's a dry region with chilly offshore waters that make for nippy nighttime lows. We were there in July, which is winter in the Southern

Hemisphere. We donned sweaters and pants as morning lows began rising out of the lower fifties.

Our first stop was the grocery store. I've loved buying groceries since I was a toddler, when my mother and I would wander the aisles of Stop and Shop or Shaw's in search of the best bargains. As Dan and I veered down through the aisles of this Chilean grocery story rapt with attention, drawn to the colorful packaging of each item. There's something soothing about the novelty of a new culture and new language overlapping with the reliable familiarity of an everyday destination.

We purchased bread, fruits, water, and snacks. It would be a long day in the mountains, far from any provisions, restaurants, or cell service. On our way out the door, Dan, who at the time didn't speak a lick of Spanish, was stopped by security guards at the front door. He was still clutching the red plastic grocery basket.

Dan shot me an inquisitive glance, unaware that the baskets couldn't be removed from the store. I could tell the security personnel were friendly, and Dan has a crackerjack sense of humor, so I decided to intervene and have some fun.

"Lo siento por mi abuelito," I said, attempting to mask my American accent. "Él tiene una tendencia a collecionar cosas rojos y brillantes." (Sorry for my grandfather; he has a tendency to collect shiny red things.)

I was grinning as the guards began laughing; Dan, in an attempt to be helpful, nodded up and down while saying "si, gracias" enthusiastically.

"No es ladrón; solamente es anciano loco," I said, my eyes twinkling mischievously. "Cualquier cosa . . . si brilla, le gusta. Hemos intentado." (He's not a thief; he's just a crazy old man. If it's shiny, he likes it . . . we've tried everything.)

The guards, amused at the dynamic and by Dan's unwavering affirmation of everything I was saying, burst into hysterical laughter. Eventually, I told Dan we had to leave the basket by the door and carry our groceries to the car.

"They were real friendly," he said. I explained why. He found it even funnier than the guards.

Then it came time to make it to our destination. The air was cool, but the sun was hot on our brows as it pierced the front windshield. We were aiming about thirty miles inland—we knew that one lone cloud would probably form near the coastline by midafternoon.

Dan had thrown out his back days before flying to Chile, and winced in pain as our car crawled along the seemingly endless dirt road; he was as hooked on seeing the eclipse as I was, though. Nothing could possibly stand in his way or hamper his spirits.

After winding through valleys and corkscrewing our way into the mountainous terrain for more than two hours, a prominent peak came into view. I checked my tattered paper map.

"We're on the two-twenty line," I called out, referring to the duration of totality we'd experience at that location. He turned and nodded. We both had the same crazy idea.

"That's the one," he said. I prepared to lug my fifty-pound suitcase up the hillside.

—∞—

The rocks slipped under my feet as I climbed the sandy, gritty mountain, wielding a suitcase that refused to cooperate. The sun's unobstructed rays beat down on my neck as I breathed in the refreshing, invigorating mountain air. I felt a foreboding sense of thrill, amazement, and anxiety building within me, knowing a once in a lifetime experience was minutes away.

"Five minutes!" Dan called out, bringing me back to the moment. The landscape was noticeably dimmer, as if lit by a dingy incandescent lamp. A light breeze dried the sweat on my face. I squinted through my blackout glasses. Only a sliver of the sun was left. An ominous presence seemed to lurk nearby.

I mentally rehearsed my choreographed camera dance one last time, ensuring I was confident I could capture the scene. Five aluminum tripods stood with heavy cameras planted atop them, telephoto lenses trained on the slender crescent sun. I knew what I saw would be emblazoned in memory for the rest of my life; the cameras were there so I could share it with others.

As I turned left to face Dan, something caught my eye. The mountains to the west no longer were visible as mounds of rusty tan: they were purple. A smoky strip arced just above the horizon. My surroundings were suddenly tougher to see. Daylight was fading fast.

"I see Venus!" Dan exclaimed, pointing at a star that had abruptly emerged from its midafternoon slumber. Night was falling before my eyes, a blue sky becoming dusky and painted by twilight within fifteen seconds. Before long, it was a color that can't be explained.

I watched as the landscape transformed into an amber-flushed painting, shadows becoming razor-thin as daylight quickly faded. I felt shivers run down my spine as the temperature abruptly dropped, the hairs on my neck standing on end. It was happening.

"Bailey's beads, bailey's beads!" I shouted. Small orbs of sunlight traced an arc like pendants on a necklace, the last perforations of sunlight beaming earthward through gaps in the rocky terrain of the moon. They congealed into a lone searing beacon, the diamond ring, which dissolved into a pinprick. Then it was gone. I removed my glasses. And that's when my jaw dropped.

The wind had gone still, flatlining in the nocturnal darkness. Insects chirped anxiously, disoriented by the unexpected nightfall. An elliptical, curved shadow had swept overhead at 10,000 mph and was parked directly above us, stars dimly twinkling in the void. The horizon was bathed in a 360 degree aquamarine.

In an instant, darkness had enveloped us, night falling as cheers erupted from the adjacent hillside. The valley floor was plunged into obscurity, glimmering stars one by one emerging from their slumber. We were in the moon's shadow. I felt the still air begin to move, the pleasant smell of wild vegetation and cactus flowers caught in the gentle breeze.

The centerpiece was the sun—or where the sun should be. Instead, it was replaced by a black hole, reminiscent of a portal to another universe. Surrounding it was the corona, the sun's atmosphere. Diaphanous tendrils of light radiated millions of miles into space, appearing like the ghostly white hair of an angel flanking the sun. The luminous wisps formed two faint loops, tracing the magnetic field of the sun. I felt as though I was on a different planet.

Standing in the moon's shadow with the solar system splayed out before you, you realize your place in life, the scale of the universe and the importance of living for the right reasons. Time seems to stand still, waves of awe, grandeur, excitement, fear, and some other indescribable emotion babbling by like gentle ocean waves lapping at a beach. You become convinced the universe is a sentient being; during an eclipse, you look it in the eye.

Until a person sees a total solar eclipse, there's no way to explain to them why people spend months' worth of savings and travel thousands of miles for something so ephemeral and fleeting. It's a transformative experience, like being momentarily transported to a different dimension. It's the universe at its most raw and elegant.

Witnessing something this incredible every two or so years is like meeting an old, wise friend; you contemplate how you've changed since your last reunion and eagerly look forward to meeting again, however brief your time together may be. Being washed over by the moon's shadow is a form of baptism, a refreshing clean slate. It grounds you.

I hope to be there for the next one, to meet this old friend again. I spend my life in tireless pursuit of those rare, beautiful moments that will remain forever etched in my memory. This one will.

—∾—

Light returned at once. The enchanted scene began to slip away like a pleasant dream upon awakening.

"The transition was way faster!" I remarked, Dan nodding in agreement. We were still catching our breath as the curtain of darkness lifted. My hands trembling, I felt like crying and laughing simultaneously. I smiled, yet my exhilaration was tainted with melancholy.

"There wasn't much of a 360 degree sunrise either," Dan said. It was true. The moon's shadow was both wider and faster than during the August 21, 2017, solar eclipse, making for a darker eclipse that built in more swiftly. I've been told in the years since that every solar eclipse has its own personality.

I climbed over a rock as my heart beat wildly, anxious for a first look at how my photos had turned out. I had invested in a massive telephoto lens for my camera; it was so bulky I had to buy a special adaptor ring to connect the lens to the tripod. If I mounted the camera by its body, the lens's weight would topple the apparatus. I clicked the camera into preview mode.

It worked! I thought, my eyes resting on the most magnificent photo I had every taken. I knew it wouldn't compare to anything NASA or a professional photographer would get, but I was elated. I had the perfect focal piece to hang as a canvas in my new apartment in Alexandria, Virginia.

—∾—

The ride back to La Serena was simultaneously peaceful and daunting. The gentle rocking of the SUV was soothing as we drove on in silence. My thoughts began to drift.

I was on a continent far from my troubles, yet I couldn't help feel that something was missing. I had just beheld the most incredible sight in the world, but without a TV camera trained on me or an audience to broadcast to, I felt incomplete. I wasn't sharing my experience with anyone. What was my purpose?

My entire life to that point had been a logical series of building blocks. First kindergarten, then elementary school, then middle school, then high school, then college. Getting into a name-brand university and traveling frequently made me feel like I was acing things so far. I had spent fifteen years as a professional student.

Now it was no longer a linear process; there was no cookie-cutter next step. The *Washington Post* was a great gig through and through, but was I making the right move?

I hadn't netted a prestigious Fortune 500 contract like most of my classmates; there weren't stock options or a sign-on bonus, and I wouldn't be having lunch on Wall Street or bragging about who I knew at Goldman Sachs.

In my heart I knew that just wasn't me, that there's more to life than a six-figure salary, but now that the eclipse was over and I was forced to confront my future, I found myself dwelling on the past four years at Harvard. Had I used them wisely?

A random clown standing in the road interrupted my quarter-life crisis. I was driving and Dan was navigating; we were reentering La Serena from the west after a three-hour drive out of the mountains. It was nice to see civilization again, the lights of convenience stores, gas stations, and city intersections welcoming us back into town. But there, stumbling through the road in front of us at a traffic light, was a real-life clown.

"Is he all right?" I said to Dan, half convinced I was overtired and hallucinating. The clown was struggling to stand upright and was continuously dropping bowling pins. He had three of them, but it seemed he could only hang onto two at once. The clown was very, very drunk.

"Oh he's *gone*," Dan said, the two of us cracking up as the clown nearly face-planted into a traffic light. He recovered at the last minute, then noticed us watching him. He tried to take a bow, but instead toppled over. Before long, I was sobbing with laughter—we were being held up by a tipsy clown.

Maybe I was doing all right for myself after all.

When the Chaser Becomes the Chased

May 22, 2020, started like any other day of storm chasing—in a $36 Kansas motel that was definitely haunted. It was 4:08 A.M., my alarm clock was rattling off its obnoxious morning clamor, and the curtains rustled in the cool, fresh breeze. After a whopping three hours of sleep, I wasn't feeling overly refreshed.

It was a small price to pay for back-to-back chase days. Twelve hours earlier, I had been dodging wind-driven hail, sidestepping funnel clouds, and watching ragged green clouds sweep overhead. A two-hour drive through blinding downpours at 1:00 A.M. brought me to Dodge City, a once-bustling gateway to the Western frontier, where I settled at the first neon-lit tenement-style lodging I encountered. The window in my room was shattered; a trash bag covered the smoke detector.

Now, I lay flat in bed, groggily prying open my laptop, and beginning to pore over data. I rolled my eyes. The setup was a game of meteorological Whac-A-Mole—so on-brand for 2020. Getting something good would be an uphill battle.

I'd been working remotely for the *Washington Post* since March, when the COVID-19 pandemic shuttered offices. Despite the inescapable doomful feeling, I rejoiced in the opportunity to ditch my cubicle. Now I could do my work in the mornings and storm chase in the afternoons.

Though overnight thunderstorms had chilled the air, it was May, and it was only a matter of hours before an insurgence of heat and humidity would turn the atmosphere into a powder keg once again. I knew that by 3:00 or 4:00 P.M. storms would be popping, and I'd have to be in position. The only question was where.

While the environment would be primed for explosive storms, there wasn't a clear-cut trigger to release that pent-up atmospheric hostility. The best bet was to follow a remnant outflow boundary, the subtle cool air exhaust of since-dead thunderstorms, south as it meandered toward the Oklahoma-Texas border.

Before I ventured south, however, I had a full workday ahead. I hastily typed out two articles, sat beneath a cigarette-smoldered wool blanket to record the day's forecast for Washington, DC–area radio, and cheered on a trio of photos as their upload trickled along. Apparently the ghoulish hotel's Wi-Fi had not reset from the overnight power outage.

By 10:00 A.M., I was fueled up, armed with a quiver of gas station rations, and barreling south on Highway 283 bound for Ardmore, Oklahoma, a quick 350 miles away. I pulled over to write another article in Englewood, Kansas, population seventy. The sky was painted a brilliant blue, with distant clouds on the eastern horizon marking the last trace of overnight storms; a stout, steamy breeze was blowing.

As I crossed the Oklahoma border and headed southeast, buffets of wind fluttered the custom-welded hail cage cantilevered above my windshield. A pair of biblical hailstorms in years prior had taught me the value of not losing my windshield during the height of a chase. Over the fan-like hum of air winnowing its way around the metal apparatus, I listened to my automated weather radar robotically recite the 11:00 A.M. weather observations. "In Woodward, it was eighty-one, wind south at seventeen, gusting to thirty-one," the voice droned through static. "And in Oklahoma City, the temperature was eighty-four degrees."

I scanned the skies obsessively, still three and a half hours from my destination. It was nearly noon. The first signs of convection, or vertical heat transfer in the atmosphere, had yet to manifest themselves; there wasn't a single puffy cloud anywhere to be seen.

I knew that something was up.

I squinted at the sky, letting the atmosphere know I was onto its trickery. And I wasn't buying it. An invisible capping inversion—a layer of warm air a mile or two above the ground that prevents air pockets from rising—was draped overhead. Like a cover atop a pot of boiling water, it suppresses and traps heat building up from below, for a while, that is. Until the cap breaks, and the atmosphere blows its lid. And that's when storms would explode.

I arrived near Ardmore, one hundred miles south of Oklahoma City, around 3:00 P.M. I was firmly in Joe Exotic country, the infamous reality star's defunct animal park just twenty minutes up the interstate from me.

I chuckled as I pulled out my phone, eager for an influx of afternoon data. My expression hardened as I reviewed the data with consternation. The outflow boundary I had been trying to line up with was directly overhead, meaning that storms would eventually develop over my location if the cap was able to break. But surface temperatures were slightly higher farther west, meaning the cap would break earlier there. As a storm chaser, you often want to be on the first storm of the day, since it can tear along unimpeded before neighboring storms erupt and compete.

I had already driven five and a half hours; what was another two more? Begrudgingly, I sighed and set my GPS for Wichita Falls, Texas. Through rolling hills I rode, traversing terrain garnished with oak trees, scrub grass, and occasional cattle farms. I passed a semitrailer abandoned in a field, the words USED PICKUPS and LOVE JESUS spray-painted on it above a phone number. I recognized it from a chase I was on in 2018. At long last, the trees began to disappear, the landscape opening up as

I approached the entanglement of highway interchanges that welcomes motorists to Wichita Falls.

I passed Carl Jr.'s, the same restaurant that had been sending me biweekly text messages since I stopped for a chicken sandwich in 2017. I parked outside an ice cream shop, cracked my truck's windows, and hopped out of my seat. The sky had a tinge of haze, a few mid-level clouds drifting overhead. The show had yet to get started. I stretched, still numb from the eight-hour drive. *This had better be worth it*, I thought.

Sauntering inside Braum's for a bowl of ice cream, I was eager to take advantage of the air conditioning. I sat, daydreaming, enjoying my lemon blueberry ice cream. After a while, I looked up. Out the window, the sky appeared clearer than it had fifteen minutes prior, the haze having disappeared.

Most people would be encouraged by the sudden sunshine, but I recognized it as a harbinger of storms. I knew it meant the cap had just broken, the air free to rise and mix out any moisture or pollutants trapped near ground level. It was go time.

I ran outdoors and climbed atop my truck, eager to get a glimpse of whatever was going on. To the untrained eye, it looked like a gorgeous day. But to my west, I spotted four or five small, cotton ball–like cumulus clouds starting to tower. One of them would become my storm. As clouds grow taller, they are subjected to changing winds with height. That can help twist them, imparting rotation and causing the storm to spin. It also affects how they are steered.

A quick visual survey told me this particular clump of clouds was about twenty miles to my west-northwest; I decided to head a few miles north. Just ten minutes later, the western horizon was dark. One of the towering clouds had swelled upward so quickly that it was blocking out the sun. Rain was beginning to fall beneath it, radar revealing a weak kidney bean–shaped patch of blue. The shower wasn't overly powerful yet, and wasn't even producing any lightning, but it was rotating.

I crossed the Red River into southwest Oklahoma, waved to the Kiowa Casino, and parked in a field just west of Grandfield. The storm was moving northeast, but I was betting on it becoming a right mover and turning. My gamble would eventually pay off. Except for the pitter-patter of sporadic errant raindrops on the windshield, it was silent. A severe thunderstorm warning was in effect, and I watched lightning flash in distance, arcing through the yellow sky from the storm's sharpening base to the ground. Curtains of rain obscured the view to the north, but I didn't care about that. I wanted to see the updraft, where no rain was falling.

The updraft is the part of a storm where warm, humid air spirals inward and upward. In rotating supercell thunderstorms, it spins like a barber's pole, and is where the most dangerous severe weather, including tornadoes, can develop. Since the air is screaming upward, no rain can fall, often leaving a clear cloud base. I knew that was exactly what I was looking at. Firmly in position, it was time to sit, wait, and observe.

Blue skies remained splayed out to the south beneath the supercell thunderstorm's anvil. A herd of curious cows joined me at the barbed wire fence running parallel to the dirt road. *I guess we're in Moo Moo Meadows*, I chuckled. I watched over the course of ten, fifteen, twenty minutes as the storm's updraft matured, with tendrils of inflow racing in from the south to feed the twirling column. All the rain and hail was remaining to the right, removed from the updraft. That was a good sign.

The storm closing in on me, I decided to move about seven miles east. It wasn't raining, and radar showed a lack of precipitation over me as I straddled the edge of the storm. Yet every few seconds, one or two lone hailstones, some the size of a half dollar, would randomly pelt the ground. That's how I knew the storm's rotation must be intensifying. The hailstones were being centrifuged and flung outward from the rotating updraft.

Meanwhile, the storm's structure was evolving into something elegant and foreboding. A wall cloud hung beneath the updraft base, a sign that a more focused region of rotation was beginning to descend closer to the ground. Dust could be seen being kicked up in the distance, an indicator of rain-cooled air from the storm's rear flank downdraft rushing toward the ground and fanning out. That process can sometimes tighten a storm's rotation. It came as no surprise to me minutes later when my phone squealed and vibrated with a wireless emergency alert: a tornado warning had been issued.

Meanwhile, the radar map was empty for three hundred miles in every direction. I silently congratulated myself for being in the right place at the right time. But my celebration was short-lived. I was brought back to the moment by an earth-shattering crack of lightning that struck into an open field a mile away. Another one flashed; then another. I recognized the lightning barrage as a sign the storm's updraft could become tornadic.

Once again, it was time to reposition east. Doing so meant threading a grid of dirt roads, made heavily of compacted clay. Most of the time, this allows for easy travel, but now the rust-colored material had become a soppy cement with the first drops of rain. I realized I had to get on the main road quickly or risk becoming stuck.

I carefully piloted my truck as it glided through a three-inch-thick slop of muddy sludge, with steering virtually impossible; the vehicle jostled to and fro with every pothole, radios and dash-mounted equipment crashing down as the turbulent motions rattled the cab. Once again on the main road, I blasted south, acutely aware that the storm would soon overtake me.

As I had learned two years ago, when chasing, you never want to fall behind a storm. Getting back ahead of it is nearly impossible, save a usually futile attempt at core-punching, which requires going through massive hailstones, extreme downpours, and damaging winds.

With every passing minute, the storm was evolving into something monstrous. I was barely three miles from the mesocyclone, or rotating part of the storm, but I couldn't see it. I was driving southeast, knowing that the momentary visual sacrifice would be worth it to position for an even better view later. I was two miles away from my intercept location as I hedged the edge of the storm, about to cross into Texas. The road curved right as I retraced my steps south over the Red River. And that's when my jaw dropped.

The supercell had morphed into a mothership, a two mile-wide cylindrical mass of roiling clouds churning just 800 feet above the ground. It resembled a thick stack of pancakes, the rotating column dark, rugged, ominous and otherworldly. It was a mass larger in volume than Mount Everest, spinning and floating, all while whipping out bolts of electricity. To the left, a razor-sharp southern edge to the mesocyclone marked the sharp transition between a pleasant spring afternoon and a storm worth running from. On the right, the mesocyclone's edge blurred against a translucent veil of heavy rain and hail.

With a storm this powerful, I had no doubt that softball-sized chunks of ice were falling just two miles to my west. I steered my truck down an off-ramp and took a right onto the first road I saw. Trees lined either side of the road, blocking the view. I cursed under my breath, but noticed a randomly placed twenty-foot pile of dirt up ahead. I stopped the truck, grabbed my camera, and climbed the anthill-like embankment of earth. Standing atop that mound of dirt, I felt like I had summited a mountain. The fruits of my obsessively produced forecast had paid off, and now I was witnessing the most spectacular storm structure I had yet seen in my life.

From the mammoth ground-scraping mesocyclone emerged a whimsical funnel below, which later was confirmed as a tornado. Sirens blared, an eerily peaceful cacophony completing the intimidating scene. Unbeknownst to me, six-inch hail was falling the equivalent of seven or eight city blocks away, challenging Texas state records. The gargantuan stones were leaving craters in the ground, and even boring through rooftops into homes.

Not in the mood to total my truck, I fled south at the last minute, noticing a few clouds bubbling in the distance. After one or two more photo stops following the initial storm east, I decided to bet on the new clouds becoming storms. It's often better to play the southern cells, since they have the best, most uninterrupted access to a supply of warm and humid air. It was a call I am glad I made, though one I wish I had made fifteen minutes sooner.

I raced east, hitting speeds of 75 mph before ducking south. Glancing at my dash-mounted radar display, I could tell that explosive thunderstorm development was once again ongoing. The storm had barely reached 50,000 feet in height when a tornado warning squealed over the radio.

"A confirmed tornado was located near Bellevue" the empty, robotic voice jabbed. I was still ten miles too far north. I had missed it. But where there is one tornado, there are often others, and I prayed the storm wasn't done producing. I exited the highway and once again continued south.

In the moments before entering every supercell thunderstorm, there's a moment of pause that washes over me. It usually comes as daylight vanishes, a few seconds after I turn on my headlights; just before the first raindrops, and just after the wind has gone still. I silence the radio, tighten my seatbelt, and lower my armrest. *Here we go again*, I think. *There's no turning back now.*

Then it hits, in this case like a car wash. The strongest storms often have the sharpest precipitation gradients. There's no gradual arrival of the heavy rain. You're either in or out. And I was in it.

My windshield wipers flailing wildly to and fro, I peaked at the GPS map and radar display guiding me. "Hmmmmmmm . . . ." I muttered. "This is going to be tight."

The storm wasn't moving fast, but it was moving east; I was approaching from the north at a right angle. That meant I'd be hard-pressed to get into the storm's vault, or precipitation-free notch east of the circulation, in time. I'd instead end up plowing through the hail core, escaping south of the rotation's path right before it passed over my location.

I considered my options. I could bail on the storm, wait for it to pass me, or thread the needle. Which meant slicing into the heart of the storm's rotating hook echo.

But threading the needle can be iffy. If there is a tornado on the ground, you might not see it until the last minute, when you emerge from the rain and hail and it's virtually on top of you. It's easy to fall into the trap of trusting radar data, but it's often outdated, and can lead to complacency. Plus, rural cellular networks make it unreliable, with spotty service at best. I didn't see any indicators of a tornado on the ground though based on radar aberrations, and decided to continue south.

The familiar high-pitched metallic pinging of hailstones ricocheting off the hood commenced, a sound I find oddly soothing; once again, I was right at home in my natural habitat. Sheets of rain still poured down, the hail becoming larger. It was no longer striking my windshield directly, now large enough that it couldn't fit through gaps in the fence-wire welded to the hail cage above. I knew that meant the hail had to be about the size of half dollars.

It was growing louder, a few golf ball–sized pieces mixing in. I reached for my safety glasses, a precaution I always take in case of shattering glass. My front and back windshields were protected, but the side windows were not. I'd never lost any of them, but there's a first time for everything.

Cresting a hill, little was visible against the sky's greyscale background, the hail shaft greedily sapping color from its surroundings. After minute or two being bombarded by hail, I began to discern the silhouette of a wall cloud about five miles to my southwest, casually lurking halfway above the ground. I was closing in. But just then, an impact struck the roof of my truck. It sounded—and felt—like a brick had been dropped on me.

I watched an icy projectile explode on the roadway up ahead, a white blur disappearing in an explosion of fragmented shards. A mischievous smile crept across my face. I knew what was coming. Seconds later, I was in a batting cage. Baseball- to softball-sized chunks of ice hurtled out of the sky at speeds topping 100 mph. Some shattered on the pavement, while others bounced upon impact and splashed up mud in adjacent fields.

I drove through a thicket of trees, the pavement slick and green from a fresh coating of shredded leaves. It smelled like Pine-Sol. Every twenty seconds or so, the truck was rocked by an enormous *thud*, hailstones striking either the rooftop or landing in the pickup bed. Some collided with the hail cage, harmlessly deflected from the flexing web of fencing and metal. A few even skimmed along the driver's side exterior, narrowly missing my window.

Eyeing the wall cloud and knowing the hail had another ten minutes to go, I decided not to chance it. I swiftly pulled off the road and backed into the driveway of an isolated farm house, praying whomever lived there would be friendly. I thought back to the gun-toting Oklahoma man who threatened to shoot me for turning around at the foot of his driveway just days earlier.

No lights were on in the single-story house, either because no one was home or because the power was out. A mound of dirt sat in the front yard, with a few dilapidated strips of chicken wire circled into short makeshift pens or trellises. A rusted green riding mower sat surrounded by clumps of grass adjacent to a bird feeder. A stone birdbath, perfectly positioned between two trees yet tilted slightly off-kiter, was overflowing with rainwater. The home was covered in white vinyl siding, a tin roof hanging over the front porch and wrapping around to form a carport occupied by two white pickup trucks. A pair of Big Wheel tricycles were strewn haphazardly about the front yard. Each thunderous impact on the home's roof sounded like someone swinging a hammer against sheet metal.

"Is it gon' make a tornado?" a voice shouted. Startled, I whipped my head right, a man in a plaid shirt pressed up against my open passenger-side window. Apparently, somebody was home after all. "It's trying," I said, pointing the wall cloud.

"Yeah, it's getting close," the man said anxiously, shaking his head. I tried to hand him a hard hat, incredulous that he would be nonchalantly standing outside as lethal meteorites of ice pounded down. He seemed unfazed.

"Yeah, I'm all right," he said, seemingly distracted.

In the most severe hailstorms, water in the cloud is divvied up into much larger hailstones, meaning fewer of them can form. That leaves a bit of space between where they fall. The man, however lucky, was playing a deadly game of dodge ball in a minefield.

I asked him if he minded me parking in his driveway.

"I don't care," he said, his attention turning to the sky once again. The wall cloud was only two or three miles to the west. "Yeah, that doesn't look good." He took off, disappearing into the house. I turned back to the wall cloud.

A moment later, the front door clacked open. The man and his wife, each carrying a small child, dashed toward the dirt pile in the front yard. He reached behind it and grabbed something—a door. They were heading into their storm cellar. I was alone; no cars passed by, and the breeze was still. Yet the wall cloud churned closer, producing occasional funnel clouds.

Amid my continued pummeling, I had the perfect view, but decided to bail south out of the hail-laden bear's cage before the wall cloud tracked overhead. A mile south, I pulled to the berm of the roadway, now watching the mesocyclone crawl just above the ground and revolve like a malevolent birthday cake. Grass and grain bowed down in waves, showing reverence to the atmosphere above. It rippled in the river of strong inflow winds feeding into the storm from the south.

Despite a few funnels shedding off the main updraft, though, it appeared to me the storm had lost its gumption. Looking west, the sky beneath the distant cloud base was orange. It was around 8:15 P.M. A textured lowering hung beneath one of the clouds, but radar didn't show much there. I didn't think much of it. To the east, the tops of faraway thunderstorms had become heaps of cotton candy, their rosy-pink paint strokes emblazoned against the graying horizon.

I rested against my truck and sighed. It was a good day, I told myself. A mothership, giant hail, and some funnels. Not bad for 2020. I casually strolled around the truck, inspecting for new dents. There were plenty to be found. "Oh, ho, ho!" I exclaimed, my eyes lighting up. "That's a bigg'un!"

I leaned down to get a closer look at a large divot impressed in the hood. There were a few more dents of similar magnitude on the driver's side door and rooftop. My father, a car aficionado, would have cried if he saw them. But I was ecstatic.

"Battle wounds," I declared, no one around to hear me but the breeze. I smiled smugly, satisfied with the scars of a good day.

I often hope that my life winds up being like my truck: beaten, driven to the limit, and with a hell of a story to tell. Some people never take their vehicles out of the garage, passing up the chance to go for a ride for fear of getting them dented or dirty. Sure, those cars will always look pristine, but their odometers are empty. My road may not be paved, but every scratch, scar, and bump is a memory, an experience, part of the journey. I want a life with a lot of dents.

Little did I know the day had more in store. A quick check of radar showed that storms were beginning to clump together and grow upscale, becoming a cluster with a diminished tornado potential. I decided to pack it up and head back to a hotel an hour and a half east in Gainesville,

Texas. I drove north, then east, the dusky sky illuminated by constant lightning strikes. It was as though monochromatic police strobes were bearing down on me.

Eventually, the storms reshaped into a line oriented southwest to northeast, with several embedded more intense cells. Wind and hail were still concerns, but the tornado threat was quickly decreasing with the setting sun. From in between flashes, a cloud to my north seemed to be dipping lower than the rest. Instinctively, I pulled over, watching out the windshield. A few specks of rain rested on the glass, while tree frogs and crickets sang their chorus outdoors.

Exhausted but relentless, I opened the RadarScope app on my phone, the display resembling a bucket of spilled paint. "Humph," I grunted. *Maybe a little something-something?* The rotation seemed fleeting, though, and I wasn't convinced it would hold together within a line of storms. I decided to instead sit back and seize the opportunity to post some photos and videos from the day to Twitter, hoping to capitalize on the ongoing storm buzz and pick up a few new followers.

Suddenly, however, my phone yelped, hissing its three-tone alert. I leapt, spilling my water and knocking my Nikon camera onto the floor. I suspected a flash flood warning had been issued. I reached for my radio.

"At 8:27 P.M. Central Daylight Time, a severe thunderstorm capable of producing a tornado was located near Bellevue, or seven miles west of Bowie," the radio warned. I sat upright, surprised. It *was* a tornado warning. And I had just been in Bellevue minutes earlier. Flipping back to radar, I could see why: rotation had increased dramatically in a kink in the line, and the circulation was just to my north by a mile or two. I knew I hadn't been imagining when I saw that suspicious cloud.

I figured I was in a fine position where I was to let the rotation pass me by. The road network wasn't great, the only roadway heading from southwest to northeast into Bowie; that was parallel to how the storms were moving, too, so as long as I stayed put, the problematic cell would miss me just to my north.

But the radio hissed again. I squinted at my phone, skeptical that the signatures I was seeing were accurate. *If this is legitimate, there are two more rotations*, I thought. It was about to get ugly.

Sometimes, weather radars get tricked by high velocities within a cloud, and range folding can plot spurious signatures. *That's what's going on*, I thought. But a new scan came in, and the winds hadn't changed. All three rotations were getting stronger.

I had to face the facts: three areas of spin, all of which could be tornadic, would pass within a few miles of me; one to the north, one directly overhead, and one to my south.

I was faced with a choice, but none of the options were optimal. I was on an east-west road, with no north-south options; the nearest intersection was in Bowie, seven miles to my east. If I tried to escape north or south, it would be fifteen or twenty minutes before I was in the clear. I didn't have that kind of time. I could try to position in between circulations, but that would be risky, too. All three mesocyclones were like atmospheric sink drains, with potentially destructive straight-line winds orbiting around the vortex. Sitting between a pair of rotational couplets meant facing off against 70–80 mph gusts on a deserted roadway bordered by flimsy power lines. Plus, additional areas of spin could form, and I'd have no shelter.

Without a viable choice, I decided my best option would be to ride out the middle circulation in Bowie. I'd be in a town with access to shelter, and roadways in all four cardinal directions in case an escape route opened up, and it was comforting to know other people would be nearby. I raced east to Bowie, greeted by sirens as soon as I entered town.

The roadways were desolate, traffic lights rolling through their sequence against the backdrop of a rain-soaked roadway. The winds were still; a light rain was falling. *Maybe this won't be so bad*, I thought.

I had ten minutes until the worst was set to arrive. But the radar wasn't encouraging—it wasn't looking good. I knew that less than a third of rotational couplets produce tornadoes, though, and there was a majority chance the spin would pass with little fanfare. I was wrong.

Winds along the thunderstorm gust front showed up within moments; so did rain. The rain progressively got heavier, winds now lightly stirring. Radar said the rotation was just overhead. No tornado. I shrugged off the storm, and used the GPS to route myself to Gainesville, Texas, to spend the night. *I guess that's it*, I thought.

As I drove north through the heart of town, I knew something was wrong. The rain was increasing in intensity, and the winds beginning to change direction. I realized that the radar had been scanning high aloft in the storm, and wasn't representing conditions at the surface; the vortex near the ground still hadn't arrived.

Flash! A bright blue burst of light lit up the landscape to my north. Then another. *Power flashes*, I thought—a sign of electrical infrastructure being damaged or destroyed by high winds.

Around that same time, a twig bounced off my window. It hadn't come from a nearby tree, though; it fell from the sky. And something had to have sucked it up there.

Suddenly, the town went dark. Main Street was pitch black, the sound of the sirens vanishing as the wind whispered louder. Rain fervently splattered on my windshield, as if trying to seek refuge inside. Then the fog hit, a milky shroud swallowing my vehicle. Visibility dropped to hardly fifteen feet, gusts of wind shaking the truck. The air was becoming saturated, but the temperature wasn't changing. There had to be a powerful funnel of low pressure nearby.

I unconsciously slowed to ten miles per hour, then five, then four, then two. With the wind blowing straight at me and leaves and debris rocketing past the windows, I thought I was still driving rapidly; in reality, I was stationary. That's when the edge of the tornado arrived.

The truck leapt side to side, the wind working as if to pry off the hail cage or the hood. Knowing the heart of the tornado was seconds away, I frantically checked my surroundings. Through the maelstrom I could faintly make out a brick wall thirty feet to my right. I revved the engine, clipped the curb, and pulled up in front of the wall. It was one of several; the structure was apparently a self-service car wash.

I unhooked my camera from the dash mount, making sure it was still recording. Shelter was close, only ten feet in front of me, but that was still too far away. Wind gusts at 80–90 mph had pinned my door shut, as if an invisible linebacker was working against me.

I kicked my legs against the center console, stretched out my body, and shunted the door ajar with my shoulder. It was my shot to escape. An instantaneous lull afforded me a second or two to slip out, suddenly bracing against the tornado's fierce winds. A spattering of raindrops struck me, stinging as if they were solid.

The truck's door slammed behind me, but I didn't care. I was already inside the wash wells, crouching in the lee of a steel- and plumbing-reinforced cinderblock divider. Assuming it wouldn't topple, my concern shifted to the corrugated roof. I watched as tree limbs, building materials, and a litter of other debris hurled by, projectile silhouettes against the headlights of my parked pickup truck. A piece of sheet metal narrowly missed the vehicle, careening south in winds gusting near 100 mph. The wind sounded like an enormous compressor.

"Alrighty guys, we're getting winds of one hundred miles per hour!" I shouted into my camcorder, pointing it outside the building. "We have debris flying by right now. We are likely in a tornado." The final words faded into the jet-like roar of the wind. But it was over as fast as it came.

Forty seconds later, the winds relaxed, the ferocious rush of air receding. It was louder in my left ear than right; I assumed the tornado was moving east-southeast. "So much for sounding like a freight train," I muttered, rolling my eyes and smirking.

I sauntered back over to the truck, pulling the hood of my rain jacket over my eyes. The air smelled like a giant lawnmower had just churned up all the dirt and trees. I plopped down in the driver's seat, turning on the air conditioner to wax away the fog on the window. I shivered, soaked in rain and sweat, and clicked on the seat heater. I yawned, suddenly aware of how tired I was.

Fighting the urge to shut my eyes, I tapped my phone, once again preparing to route myself to Gainesville, Texas. I frowned. The network was down, probably due to severed power or toppled cell towers. I knew I had to head north. I shifted into reverse, checking my backup camera for debris. Specks of shrapnel glinted in the headlights. I made it fifty feet to the main road before I was forced to stop. A pair of tree limbs, each a foot thick, were blocking the roadway. *I was just here*, I thought. Traffic signs, branches, and occasional building supplies were strewn about, slowing my progress to a crawl. Toppled wires blocked multiple lanes. I relied on my headlights to help me avoid ongoing flash flooding; pools of water a foot and a half deep inundated low-lying parts of the roadway.

I decided to tour a neighborhood that looked to have been in the apparent tornado's path. Widespread EF0 to EF1 damage, commensurate with winds of 80–100 mph, was evident. Virtually every yard seemed to have trees snapped or downed, some resting on homes or having fallen on cars, while sheds and carports lay mangled. The next morning, the National Weather Service in Fort Worth confirmed it was an EF1 twister that had stampeded through town.

Meanwhile, it was now past midnight. I had been up for twenty hours straight, driven nearly six hundred miles, been to three states, and been concentrating nonstop since around lunchtime. It was time to get to the hotel.

White-knuckling it the remaining ninety minutes to Gainesville wasn't fun. In heading east, I caught up to the storms, reentering their heavy rain from behind. I drove on the crown of the road, avoiding dozens of power lines that had been knocked over like dominos in bursts of straight-line winds. When I finally arrived to my hotel, the winds were back at 60 mph. I parked in the parking lot, slung my backpack over my shoulder, and nonchalantly shuffled into the hotel, refusing to acknowledge the storm again raging around me.

"Wonderful weather we're having," I chuckled, exchanging a smile with the hotel clerk. A man and two children, presumably her family, were piling towels in the lobby where the ceiling had sprung a serious leak. I tiptoed around the newly formed indoor swimming pool, lugged my back down the hallway, and fumbled as I dug around in my pocket for the room key.

A swipe and a double beep later, the door swung open. The familiar smell of must, mothballs, and cigarette smoke greeted me as I collapsed down on the bed. I signed contentedly, smiling, thinking about my truck. "Today, that was a good dent," I said.

The Never-Ending Hurricane Season of 2020

Significant tornadoes were entirely absent from the Great Plains during May 2020, providing beleaguered residents of tornado country reeling from COVID-19 a window of relief. Mother Nature's prophetic quiescence was just to lull the weather weary into a false sense of security, however; an early start to the Atlantic hurricane season came with the formation of Tropical Storm Arthur east of Florida on May 16.

Hurricane Bertha came together offshore of the Carolinas eleven days later. By the books, the Atlantic hurricane season doesn't begin until June 1, though the National Oceanic and Atmospheric Administration (NOAA) is mulling shifting up the start date to account for climate-driven trends.

The 2020 Atlantic hurricane season ended up as the busiest on record, featuring a record-shattering thirty named storms, two and a half times the yearly average. Thirteen of them made landfall on US soil, with virtually the entire Eastern Seaboard and Gulf Coast included in tropical watches or warnings at some point during the season. The National Hurricane Center burned through its preset list of hurricane names, forcing meteorologists to dip into the Greek alphabet.

The season featured a series of astonishing feats, including back-to-back Category 4s that struck the same parts of Nicaragua and Honduras less than two weeks apart. Stateside, Hurricane Laura was the cardinal storm of the year, mowing down forested swaths of southwest Louisiana as the most formidable storm to strike the Bayou State since 1856. Winds gusting to 150 mph expunged structures from the coastline.

I rode out the storm in the Center for Severe Weather Research's Doppler on Wheels (DOW), a mobile Doppler radar truck. Anticipated storm surge limited where the DOW could be parked and its antenna erected, relegating Josh Wurman and Karen Kosiba, the center's lead scientists, to settle on a bridge over the Sabine River on the Texas border. We got winds to 111 mph, but missed out on the eye.

I'd have a do-over in mid-September, when Hurricane Sally ambled into Alabama as a high-end Category 2 storm. Like virtually every other landfalling storm in 2020, it trumped expectations as a borderline major hurricane upon landfall. Just twenty-four hours earlier, it had been a tropical storm struggling to persevere amidst modest but immutable wind shear.

My bosses at the *Washington Post* made the call at 2:41 P.M. on Monday, September 14: I'd been approved to fly down to the Gulf Coast and cover Hurricane Sally from the field. I'd been persistent (and perhaps annoying) in my badgering, arguing that the HMON, an especially punchy hurricane model, would prove correct in its projections for Sally to move ashore as a trenchant, dangerous storm. A recent flare-up of fifty-thousand-foot-tall thunderstorms near its center helped my case. Jason called me with the news while I was recording my afternoon radio hits.

Thirty minutes later, I was out the door, armed with a trash bag stuffed with clothes, my camera bag, and a pair of hurricane goggles. I knew I wouldn't sleep much over the next forty-eight hours, but I was ready. My Delta flight took off from Reagan National Airport at 4:40 P.M. The sky was crystal clear.

—w—

I landed in New Orleans at 8:00 P.M., picked up my rental car, and zipped east to Biloxi, Mississippi.

"Where you heading?" the counter clerk at Avis asked. A two-second awkward silence ensued as I racked my mind to figure out if he was making small talk or about to rescind my rental.

"Just making a quick work trip," I replied, smiling jovially and doing my best to feign ignorance. "I love when I get to come down here. The food is so good." If there's one thing I'd learned in my travels, it's that folks from the South love to talk about their favorite styles of cooking. Within ten seconds, the desk agent had completely forgotten about the soon-to-be hurricane I'd be driving into.

I shacked up at the Star Inn, a two-star hotel on the water. JACUZZI ROOMS, a sign in front of the premises read. Red corrugated panels covered the gabled roof. *That's good construction*, I thought.

Sharkheads, a family destination and gift store, was across the street. The hot-pink building was propped up atop stilts. I was officially in hurricane territory.

Unfortunately, my room didn't have a Jacuzzi. That was probably a good thing, since I didn't necessarily want to take a bath where, statistically, a classroom's worth of children might have been conceived. Instead, I recorded two quick video forecasts for social media, sent them to Jason at the *Washington Post*, and slipped into bed. While the Twitter posts weren't required, each snippet I uploaded was gaining a lot of traction and growing my following. Maybe prospective TV employers would take notice.

Only a gentle, humid breeze stirred outdoors, the air thick with moisture but devoid of any rain.

~

I awoke Tuesday morning at 6:00 A.M. and immediately reached for my iPad. Data revealed that little had changed overnight; Biloxi was on the knife-edged fringe of Hurricane Sally's precipitation shield, and despite threatening skies and dark clouds, the grass was dry. A few specks of blue sky smirked in the west, with doomfully dark clouds banked low in the east. After a quick morning article, I drove east at 9:00 A.M., a midmorning dusk adhered to the horizon.

The roadway was damp as I crossed the border into Alabama, light drizzles becoming moderate downpours as I traversed a steep gradient in rainfall intensity. Storm surge was already beginning to slash the coastline of Mobile Bay. I circumnavigated to the east, holing up in Gulf Shores, Alabama, by noontime.

Puddles transitioned into lagoons as I approached my destination, a foot or two of water gushing down roadways near the shoreline. Sheets of rain were pouring down by lunchtime, a peculiar rain that somehow was both warm and seemed to extinguish my corporeal pilot light. Gale-force winds swept through the streets, causing stop signs to flutter as if agitated. I stopped at a Circle K to stock up on potato chips, mini muffins, water, and other final essentials.

"Nap time," I said, eager to close my eyes for just a few moments. Fourteen minutes later, I was conscious again, recharged and ready to tackle whatever Sally was about to throw my way. My iPhone shot me an angry look, an aggressive dialogue box reading LIQUID HAS BEEN DETECTED IN THE LIGHTNING CONNECTOR appearing on-screen. I sighed and crammed the phone in between flaps in the air conditioning vent, praying the 38 percent charge it had would hold through the evening. Road signs were beginning to quaver in the building winds.

I went to shift into drive but, just then, my phone rang. It was Jason.

"Hey there," I said. I hoped that wetness in the microphone wasn't garbling my speech.

"Hey Matthew. How are things going? I wanted to check in and see your status," he replied. I knew he'd be putting in long hours, too. As it was, a typical workweek for him usually ran sixty hours. I bet he was well on his way to ninety.

"I'm in Gulf Shores, but I'm a bit worried about the Bay," I said, referring to Mobile Bay. I was on the eastern side. If Sally wound up wobbling farther west and I had to reposition closer to Dauphin Island, just twenty-five miles away, it'd be a circuitous three-hour trip. Jason reassured me that the forecast was on track.

"Are you in position to do a couple feeds?" he asked. Feeds are two-paragraph blurbs that can be dropped into a live update file, which is the default format for rapidly unfolding breaking news events. They had become useful in our weather coverage when high-impact storm systems were targeting a populous region.

I drove down Highway 59 over Portage Creek, the unofficial threshold splitting the mainland from the barrier islands. FIRST BAPTIST CHURCH WELCOMES YOU was tacked onto a whitewashed building in blue ceramic lettering. The vegetation was comprised mainly of pine and palm.

The ocean was inundating parking lots and lapping at the edge of the roadways, the plashy ponds merging with canals and rivers. Water was swallowing mailboxes and staircases, knocking on the front doors of homes and businesses. Breezy squalls with gusts over 50 mph were pivoting into the region. Things were escalating quickly. I filmed a quick report on Windmill Ridge Road. Water was cresting over my ankles.

By nightfall, the surge was becoming ominous. Instead of just a submerged shoulder or blocked lane of traffic, the best case-scenario meant driving on the crown of the roadway through half a foot of water. I'd saved money at Avis by opting for a red compact car that was basically a glorified golf cart. I couldn't help but instinctively lift myself off my seat every time I drove uphill.

Traffic lights were blinking as darkness fell on Gulf Shores, the entire city abruptly vanishing as power cut out. The only other cars on the road belonged to police. I headed two miles inland to the Microtel by Wyndham, making sure I was off the barrier island as the 8:00 P.M. curfew neared.

Four to fourteen inches of water awaited me as I rounded the corner into my hotel. I parked my rental atop a raised concrete walkway. That put the lawnmower-esque automobile about eight inches over the asphalt.

Around that same time, ever-capricious Sally started doing something it wasn't supposed to do: strengthen. Its laggard speed had been churning up cooler waters from deeper in the ocean, but the storm didn't seem to care. Doppler radar revealed winds ramping up in Sally's sharpening eyewall, while satellite showed a circular void appearing as the eye emerged.

I was just forty miles north of Sally's center. The weather wasn't great, but it paled in comparison with what was to come. Winds were gusting 40–50 mph, but only in irregular bursts. Lights flickered in the hotel and across the surrounding neighborhood, a pulsating beep emanating from the hallway each time electricity came back.

The power went out for good around 10:30 P.M., coming simultaneously with a lull in the winds. This was the moat. All the air rapidly rising in the eyewall had to sink somewhere. Some subsided in calm bands that radiated outward from the center. They were brief moments to collect oneself before the next round came in.

The combination of wind and rain made me feel like I was inside a washing machine. I also noticed a few couplets of rotation on radar passing just to my north as Sally's changing winds with height tried to spin up sporadic tornadoes.

Then, around 11:00 P.M., things got ugly. Even though Sally was barely moving, the northern edge of its eyewall had arrived. Spasmodic gusts became a constant, sustained roar, with entire foot-thick trees swinging and branches whipping to and fro. A series of incandescent blue power flashes lit up the sky in rapid succession, the result of blown transformers or severed utility lines.

My cell phone blared with a flash flood warning. Close to ten inches of rain had already fallen, and another foot-plus was on the way. At midnight, Sally was declared a high-end Category 2 storm with 100 mph winds. Then it strengthened even more. The door to my hotel room rattled on its hinges predatorily as the breeze outside sent air-pressure fluctuations throughout the building.

Even though the outer eyewall was scraping us, Sally was only moving at 3 or 4 mph. I'd have hours before the worst arrived. I forced myself to get a few hours of shut-eye, knowing Sally's onslaught and damage would spell a long day ahead. My eyelids were heavy, and the hotel room pitch black; I noticed no difference when I closed my stinging eyes.

I awoke to the sound of crashes. My phone clock read 3:32 A.M. The wind was screaming, the roar injected by what sounded like shattering pottery. A pressure washer of wind and rain was exploding on the building. Not a light was visible anywhere, but the silhouettes of trees could be seen flailing wildly, as if frenetically trying to get someone's attention. A howl could be heard in the hallway—the thunderous voice of Hurricane Sally.

Jarring jabs of wind lunged at the hotel in trembling fits. The building was shuddering in the throbbing wind. *We've got to be getting gusts over a hundred*, I thought.

Suddenly, the wind relaxed some, as if shushing me in order to whisper a secret. I listened intently. Moments later, it pulsated back even stronger than before. Over the following fifteen minutes it gradually trailed off—not smoothly, but in a jagged stair-step pattern with abrupt slackening of the furious cyclone that had tortured us for hours.

It was 4:00 A.M. and I was in the eye: a self-contained reprieve of calm surrounds by meteorological hostility. Stepping outside, I could feel the air stir against my face. Soon it was as still as death. I felt invigorated.

Tree frogs croaked and crickets began chirping, as if gradually realizing they were allowed to. They joined together in an ignorant chorus, unaware that they had only reached the halfway point. The sky overhead remained a misty, inscrutable sheath, with occasional dark patches where sinking air had eroded some of the low-level cloud cover. The serene scene belied the rage that surrounded me just miles away in all directions. The air smelled sweet and fresh, as if cleansed by a summer downpour and replaced with the tang of a wilted bouquet.

My phone's flashlight was too dim to spotlight the landscape, but I had an external broadcast lamp that did the trick. Clicking it on revealed that I'd made the right call lumbering my rental car onto the raised concrete: two feet of rain had fallen, and the parking lot was a swamp. Realizing I'd never keep my sneakers dry, I sloshed through the lot to inspect a pair of fallen trees.

"Agh!" I shouted, leaping out of the water and desperately grasping at the air. Something had brushed against my ankles! With a start, I realized shrimp were swimming freely in the parking lot. They had somehow made it miles inland. Floating in the water were shingles and roofing tiles.

It didn't take me long to find the source of the clamor that had awoken me: a forty-foot tree had been uprooted outside my window. Its root balls greeted me at eye level. Another fourteen-inch-thick tree trunk nearby was snapped. One of the hotel's doors had been ripped from the wall and thrown inside the building; the other was shattered, punctured by a two-by-four.

Some of the waters in the parking lot began to recede after approximately ninety minutes in the eye. I glanced at my phone's radar display—the eye should linger overhead through sunrise, but a second breath of wind, albeit not as intense, would accompany the backside.

For the time being, I stood outside, wondering what was happening a block, a mile, ten miles away, how others had fared riding out the storm. Traffic lights and utility poles were downed virtually everywhere, with gas station canopies mangled into industrial skeletons. No one would see that until morning, however. Instead, I was looking skyward, unsuccessfully searching for stars through the veil of overcast.

The Big Red Lawnmower was one of two rental cars I'd damage within a month. While leaping it onto the cement sidewalk had saved it from a watery grave (two vehicles in the parking lot were totaled), I'd still scuffed up the plastic underside. Four weeks later, I'd wreck the driver's side door of a RAV4 in Louisiana after foolishly opening it amid strong winds. (The two flat tires I also popped circumnavigating downed power lines were icing on the cake.)

That time came during a trip back to the Gulf to cover Hurricane Delta, which hit southwest Louisiana as a Category 2. The city of Lake Charles was a sea of blue tarps replacing roofs for the nearly half of structures that had lost them to Hurricane Laura's 140 mph gusts in August. Delta made landfall in October, an unfortunate double whammy.

There's a firm link between human-induced climate change and the growing intensity of hurricanes, a correlation that was brought to international attention following the devastating 2020 hurricane season. The oceans are absorbing about 90 percent of the excess heat tied to greenhouse gas emissions, warming, and yielding more dangerous storms.

Ten storms during 2020 underwent rapid intensification in the Atlantic, tying a record last set in 1995. Virtually every storm nowadays seems to overachieve and exceed expectations. Kerry Emanuel, an atmospheric scientist and my former professor at MIT, authored a study in 2017 painting a disconcerting picture of the future.

At present, a storm will lurch 75 mph or more in intensity during the final twenty-four hours before landfall only once per century; that rate of extreme rapid intensification is very rare. Emanuel concluded that such a pace of rapid intensification may occur every five or ten years by the year 2100.

Even more alarming was this nugget he included in his write-up: one-day 115 mph leaps in intensity, which Emanuel described as "essentially nonexistent" in today's current reality, could become physically possible in the next eighty years.

Rapidly intensifying storms are a planning nightmare for emergency managers and a troublesome bother for forecasters, especially in marginal circumstances. A storm that jumps a category or two in strength before drifting ashore could spell disaster, especially in a major urban area where evacuations aren't always possible. Rising sea levels also make for more damaging surge.

Tropical cyclones are also becoming wetter, bringing major flooding farther inland. For every degree the air temperature increases, the atmosphere can hold about 4 percent more water. In continually replenished air masses, like those twirled into hurricanes, that can quickly translate to 15 or 20 percent more rain. Hurricane Harvey dropped 60.58 inches in Texas in 2017, with another 44.29 inches falling from the remnants of Tropical Storm Imelda in September 2019. Both episodes brought catastrophic flooding. Both North and South Carolina set rainfall records, too, in 2018, from Florence, with 35.93 and 23.63 inches, respectively. Hurricane Barry set an Arkansas record with nearly 17 inches in 2019.

As if that all wasn't bad enough, hurricanes are wandering farther outside their typical territories as waters warm and become hospitable for them. The latitude of maximum hurricane intensity in the Atlantic is moving farther northward, making the United States more susceptible to major hurricane landfalls. In September 2019 Hurricane Lorenzo became the farthest northeast Category 5 on record, six hundred miles from the previous farthest east Category 5s: Hurricanes Isabel and Hugo.

The tropics may not simmer for years to come, the Atlantic engrossed by a multidecade enhancement in activity. Higher-end storms will cost more insurance dollars annually, and the economic, and human, tolls will only grow with time.

Wild Kingdom

Most people know the Midwest as America's breadbasket, with endless crops stretching as far as the eye can see. But few mention what's hiding inside those crops: animals. And they all have a personal vendetta against my truck.

It all started back in 2017, when a rabbit and my truck rendezvoused in an unfortunate encounter following a hailstorm in Perryton, Texas. Since then, animals far and wide across the Great Plains have been working to avenge the premature death of Perryton's very own Peter Cottontail, targeting my vehicle in an effort to get me on PETA's blacklist.

The year 2020 proved especially problematic for animals, a snake and two birds falling victim to my travels by mid-May. Things got worse on one particular evening, when I was driving home from a six-hundred-mile storm chase bust that didn't even feature rain. Naturally, I had cued up the somber theme of Disney's *Up* on my iPod. I was an hour out from my hotel in Manhattan, Kansas, as I drove along the highway at 75 mph. A deep azure twilight hung above. Up ahead, I noticed a small shape moving in a cornfield. It was a raccoon.

"Awe," I smiled halfheartedly. "That looks like Noodles!" I said, reminded of my dog. I double-beeped a cordial greeting to what I assumed was a friendly raccoon, hoping to shoo him away from the road.

Instead, the creature catapulted itself onto the roadway, followed a split second later by a god-awful crunch beneath the right side of the vehicle. I instantly burst into tears, *Up*'s "Married Life" still blissfully pattering away. With one hand on the wheel, I adjusted my contact lenses, which were now sliding about my watering eyes. It was a fitting end to a rotten day. I said a silent prayer for the raccoon's progeny, innocently wondering if it had a family. *Do raccoons have houses?* I wondered.

That's when Mother Nature decided to go double or nothing. Out of nowhere, a second raccoon flew onto the highway as if launched from a cannon, landing immediately in front of the driver's-side tire. I ducked. The vehicle shuddered from the unavoidable impact, which was simultaneously met by my blubbering stifled sob. A string quartet rendition of Katy Perry's "Unconditionally" poured through the speakers. I choked back a wail, wondering what cruel sitcom I was in.

If I can kill two raccoons with one truck, does this mean I could kill two birds with one stone? I thought.

I later learned that the airborne raccoons hadn't only left a pair of dents—they had also obliterated my front turn signals. And as for how many meteorologists it takes to change a light bulb, well . . . apparently it's more than one, because I didn't have the tools to do it. I spent the next week signaling turns with my hand out the window.

A week later, I was in Brookings, South Dakota, returning from a soothing evening enjoying distant heat lightning flickering across the sky. It was 11:00 P.M., but the air was still warm, filled with the songs of crickets. With no one around, I rode with my windows open, enjoying the still, dry air. I signaled to the empty landscape that I was about to turn right onto the highway.

Just then, a brown blur careened into the roadway, charging at my truck as I slammed on the brakes. "AAAAAGGHH," I shouted, instinctively suspecting it was a giant, tan raccoon. It never crossed my mind those don't exist.

I gripped the wheel as a heavy object pelted the hood, followed by a dull *thud*. By now, the truck had screeched to a halt, a tumbleweed of hooves and antlers and fur somersaulting ahead. Seconds later, the deer came to a rest about fifteen feet in front of me. It shakily stood up, collecting itself the same way I do when I fall down the stairs in front of an audience. I shifted into park and hopped out of the vehicle, standing beside my truck. We exchanged eye contact. "I'm going to need your insurance paperwork!" I shouted at the deer, which quickly took its cue and scurried off.

"Look both ways next time like your mother taught you!" I yelled into the night, sincerely hoping the deer would listen. I wondered if by now I had checked off enough species to get bingo.

I thought back to my uncle, a rough and rugged Mainer who vociferously brags about hunting down deer with a rifle. I had managed to do it at 23 mph with a Honda Ridgeline. (I still think hunting would be much fairer if the deer had guns, too.) With the front end of my truck smooshed in, headlights shattered, and grill covered in brown hair, I spent the next two weeks avoiding the evening animal rush hour like the plague. It worked until my last night in South Dakota, when heavy squalls flooded fields, forcing thousands of leopard frogs—my favorite animal—onto the roadway. I rescued several dozen of them amidst a heavy lightning storm, relocating them off the highway south of Aberdeen. I helped along a couple turtles, too. Even though I was one zoo in debt, I hoped I could at least win the amphibians' popular vote.

When I returned to the East Coast the next week, I set up an appointment with a repair shop to have the deer and raccoon damage remedied. Days later, I walked into the parking garage to find the entire front bumper of the truck hanging off. Apparently, another driver, who signed his insurance note "Kevin," had misjudged the dimensions of his parking space by about half the length of a school bus.

It's like that sometimes.

It made me wonder, however, what do animals actually do during foul weather? I'd heard about elephants and other mammals behaving oddly and rushing inland ahead of the December 26, 2004, Sumatra, Indonesia earthquake. They sensed the *p* waves, or preliminary waves that preceded the quake's violent surface shaking, and hightailed it inland before the deadly tsunami arrived. More than 225,000 people perished in the tidal surge that followed. I was curious if animals had meteorological instincts, too.

The more I dug into it, the more I realized they were just as subject to the elements as we humans were. Following the Canton, Texas, tornadoes on April 29, 2017, I encountered multiple dead birds on the ground. At the time I was horrified, but it was an experience I would regrettably repeat multiple times in the future in the aftermath of significant hailstorms.

It made me think back to jackrabbits I'd seen outrunning tornadic storms, or the times I'd found evidence of birds crowding in the eyes of hurricanes. During Hurricane Harvey's Category 4 assault on Rockport, Texas, in 2017, a blue splotch in the eye demarcated jagged or irregular shapes evaluated by radar. It turned out to be birds that were gradually siphoned into the eye, unable to escape the storm's inward pull. The same thing happened with Hurricane Sally in 2020.

During the infamous hurricane of 1938, which lashed southern New England with winds gusting up to 186 mph, yellow-billed tropical birds were recovered in Vermont for the first time on record. A Cory's Shearwater was found a week after landfall in Peru, Massachusetts, and both a Great Shearwater and Sooty Tern were recovered in Massachusetts in the coming days. It's unclear if the birds were ever able to return to their home ecosystems.

When Things Don't Go Quite Right

I may lead a charmed life, but that doesn't mean things always go exactly according to plan. In fact, oftentimes things unravel exactly opposite to how I had hoped.

It's true in all walks of life: personally, professionally, and meteorologically. Rejection is my middle name. I still get about a dozen emails per month that begin with "we regret to inform you." It can feel like being eternally left-swiped on Tinder.

It took about ninety emails to various newspapers starting when I was fifteen before a single one on Cape Cod gave me the opportunity to write. The repudiation of my abilities only grew with graduation. I began to wonder if I wasn't cut out for meteorology.

In retrospect, each soul-crushing rejection was a covert blessing. The universe had a plan, and it wasn't for me to end up in Appleton, Wisconsin, or Omaha, Nebraska. Each rejection was life's way of helping me to find "home" sooner than I had expected. I learned to sometimes just trust the process and have faith; even still, it isn't easy.

Sometimes the atmosphere delivers equally swift blows of diffidence; each dose of humility, however, can cost hundreds or thousands of dollars. Storm chasing is an expensive hobby (or, if you're me, way of life), and a busted forecast translates to hefty sums of gas money, hotel fees, tolls, and driving time down the drain.

The year 2020 was disappointing in all respects. It began with the COVID-19 pandemic, was punctuated by a five week-long storm chase that yielded no tornadoes other than the one that hit me, and ended with an ill-fated trip to Chile. There was no end-of-year redemption.

I had hoped to fly to South America for a total solar eclipse on December 14. Like the one sixteen months prior, it would cast a shadow over a sliver of Chile and Argentina for a little more than two minutes. The path of totality was to fall about six hundred miles south of its predecessor's. Instead of the Chilean deserts and prairie, however, totality this time would be visible from Aruacania, a region known for its mountains, lakes, and volcanoes.

Climatology favored Argentina as the most promising place to view the eclipse, with the best shot of clear skies east of the Andes Mountains. The county's width would also afford better chances to reposition and dodge cloud cover compared to Chile, which only averages 110 miles between the Pacific and its eastern border. Argentina was shuttered to foreigners during the pandemic, however. That meant Chile was my only option.

International travel was virtually nonexistent in the weeks leading up to my trip. I didn't know anyone who was flying anywhere, never mind out of the county, and the Chilean government was a moving target as policies, travel restrictions, and commune-level *paso a paso* phases were constantly adjusted. I spent weeks drafting a color-coded plan, backup plan and half a dozen other contingency plans. In the end, I had to choose Plan G, which I had shaded in magenta.

It was announced two days before my trip that Santiago, home to six million people and the nation's main international airport, would be regressing to level two out of four *transición* state. Since interregional travel was only permitted between municipalities in phases three or four, I had to scrap my plans to remain in Santiago for two days, as well as fly to Calama in the Atacama Desert, the driest nonpolar desert on

Earth. Instead, I arrived and zipped directly to Pucón on the shores of Lake Villarrica, a nine-hour drive south, and holed up in a cozy Airbnb.

Showers of rain and ice pellets persisted for days before the eclipse, with temperatures in the upper forties and lower fifties. I gingerly balanced logs in a woodstove, stoking it around the clock to keep warm. While the rat-a-tat of nonstop rain on the cabin's tin roof was aesthetically soothing, it embodied my worst fears: that the eclipse would be spoiled by rain. An atmospheric river, or narrow filament of deep tropical moisture, was draped overhead. Its departure looked to coincide with the arrival of the eclipse. With little else to do, I apprehensively toured a volcano and trekked toward its summit until the cold air made me lose feeling in my lips.

When the morning of eclipse day finally came around, I woke up at 2:00 A.M. to overcast skies and light sprinkles. Weather models indicated the coastline would have better odds of rifts in the cloud cover, so I spent five hours driving through pastures and on dirt roads there. The eclipse occurred around lunchtime, the sun's sullied, diffuse glow through the cloud deck suddenly disappearing as a misty darkness briefly enveloped the cloud-covered landscape. There was nothing to see otherwise, the sun's divine corona effectively extinct.

After flying eight thousand miles, spending $2,000, and navigating to a horse farm in the middle of nowhere on a different continent, all to see nothing, I sat in a field and cried. My tears camouflaged with the raindrops that had collected on my cheeks. Skies cleared as the back edge of the atmospheric river exited half an hour later. You can't win them all.

Failed chases are part of the job. Missing big events, even if by a few minutes or miles, make the victorious chases sweeter. It's like life—hurdles are what make us appreciate when things go right. I knew that, inevitably, things would go right again.

~

Some of my failures have been intentional, like on March 25, 2021, when I chased a high risk tornado outbreak in Alabama. Those red-letter severe weather days are the ones that spur school closures and early dismissals from work. Families hunker down close to TV sets and radios. Every rustling leaf is a voice of unease. My initial goal had been to document a tornado from close range, but ultimately staying alive trumped that aspiration.

I spent the afternoon in Tuscaloosa, Alabama, a city that had been ravaged by an EF4 on April 27, 2011. Around 4:18 P.M. I was in Moundville, a town of 2,500, just to the south.

Sirens were blaring a pair of tornadic supercells pushed northeast. I was waiting for the northern one. It quickly died as the southern supercell went into beast mode. The Storm Prediction Center warned that "a corridor for a long track intense to potentially violent tornado [was] expected."

Screaming south, I knew the odds were against me making it into position in time. It was the eeriest tornadic signature I had ever seen on radar during one of my chases, the unimposing pixels on my iPhone representing a mile-wide monster buried in rain. The map smelled of invisible death.

There was no hail or wind as I drove south into the storm, passing through dense forests in hilly terrain that reminded me of New Hampshire. I was aiming for the town of Greensboro. Rain was falling as I entered the supercell from the north, but there was no hint of what was lurking four miles away. The enormous tornado was barreling toward the road I was driving on. Eventually I reached a geographic point of no return.

I can do this, I thought. *Three minutes. We've got this.* That was how long I'd have before the tornado crossed my location if I proceeded. I could probably do it, but any slowdown, road closure, or tree in the road would leave me with no escape route. This wasn't Oklahoma, where a map of the gridded road network could be used as graph paper. I could go either forward or backward, and that was it. There were no other vehicles to be seen. I foraged onward.

As I drove, the tiny blue icon marking my location seemed to be on a crash course with the gargantuan ball of debris overlaid on the radar. I was no longer chasing: I was tempting fate. Sighing, I lifted my foot from the accelerator as Ridgie came to rest in the roadway. *Not today*, I thought.

What happened next still gives me chills. After about four minutes, the sky suddenly darkened. Not just a little bit, but as if it was on a dimmer and had entered sleep mode. The winds went still as an indescribable blackness silently hewed past me, bringing a forced repose as it permanent engraved the landscape. Forty seconds later I was back in an ordinary rainstorm. And I'd live to chase another day.

Every chaser learns to abide by the words of Kenny Rogers in his famous song: "You've got to know when to hold 'em. When to fold 'em. Know when to walk away. And [most importantly] know when to run." The atmosphere is a source of bountiful beauty, but is a creature that deserves respect. After all, it's in charge; it always wins. We just have to remember that we're merely the guests.

A Quest to Find the Northern Lights

I was so frustrated I could scream. I was missing the tornadoes. If I wasn't buckled into seat 31D on a flight from Washington, DC, to Seattle, I'd be in the Texas Panhandle, pulled to the shoulder of some random dirt road watching a whirling mass of vaporous black cross the highway in front of me.

Instead, I was dejectedly scrolling through Twitter, irritably mumbling about debris balls and hook echoes as the plane slowly rolled down a taxiway.

"There will be other tornadoes," Allen softly reassured, seated to my right next to the window. I shot him a dirty look. I knew he was right, but sitting out an event has never been my specialty. *At least we're going on vacation*, I thought.

It was the first big trip I was bringing Allen on. He was five months older than me, graduated my same year, and, despite the fact that he went to Yale and was supposed to be my archnemesis, we had quickly become inseparable friends. Good adventure buddies are hard to come by, and Allen had already passed with flying colors.

Allen was a statistics major in school, but his true love had always been music. From an early age he quickly developed a knack for the piano, and by middle school he was a borderline prodigy. That passion expanded to the violin and clarinet, and his high school years were spent traveling the country and performing in regional and national competitions. College would take him across the world with the Yale Symphony Orchestra. For now, his day job was as a consultant at a major (boring) firm, which meant he could take his job anywhere.

Careening down the runway, I couldn't help but chuckle. This trip was a last-minute adventure born from equal parts intrepidness and stubbornness. Two weeks prior, I had jokingly asked Allen if he felt like flying to Fairbanks, Alaska to hunt for the northern lights. He called my bluff and bought a ticket. After egging each other on, we were on our way to the Arctic tundra, where we planned to cross off something that had been on each of our bucket lists since childhood.

I had timed our planned visit to coincide with the peak of something called the Russell-McPherron effect. A little-known secret to forecasting space weather, it predisposes Earth to vibrant displays of the aurora borealis, or northern lights, around the time of the spring and fall equinoxes, roughly March 21 and September 21, respectively. I had surplus American Airlines miles, was stir crazy because the pandemic, and was in desperate need of an adventure. It seemed like the logical thing to do.

Things had seemed to be coming together for the spur-of-the-moment trip—until several days before, when I inexplicably came down with a 104 degree fever, chills, and all the symptoms of COVID-19. My tonsils became inflamed and I could barely hold my head up. Both a rapid test and a PCR test ruled out COVID-19 as a cause, though, and the doctors told me it wasn't strep throat either.

The true culprit was a thermometer-related injury and illness. Days prior, I had bolted from my couch to the window to catch a glimpse of a rare cloud, only to trip over my coffee table and slice my hand open on a glass thermometer that toppled off the TV stand and shattered. It turns out that a shard had become lodged in my middle finger and stirred about a nasty infection. Antibiotics cured it, but didn't stop my sister from teasing me mercilessly.

By the time Allen and I touched down in Seattle, I was fully immersed in vacation mode. I had managed to go a full five hours without checking my email, and was more interested instead in data on the Earth's magnetic field. Darkness had settled in as we boarded our connecting flight to Fairbanks, and the numbers I was seeing suggested a low-end shot at seeing some pastel green auroral hues from the airplane.

About two hours into the flight, a faint glow emerged over the horizon. I squinted, barely able to perceive hints of washed-out green, as if looking at a faded photograph. The monochromatic arc was unimpressive, but the photos I was snapping out the window resembled something off a postcard. *Is this it?* I wondered, starting to think the northern lights may just be a photography scam. I wasn't optimistic when we touched down in Fairbanks at 1:00 A.M.

Allen and I collected our belongings at baggage claim, and I called the hotel to request an airport shuttle. We stood near the exit door, the snow-covered landscape reflecting a sliver of moonlight. Allen huddled next to the metal wall heater while I stepped outside into the tranquil night for some fresh air.

My eyes instantly stung; the cold lapped at my face, freezing the moisture in my nostrils. It felt like someone had dunked my Waffle House sweatshirt in frigid water, the bitter air permeating any gaps in my clothing. My hands trembling, I clumsily looked up the current temperature in Fairbanks. Zero degrees. I marched back inside.

"I'd like to report that there is no temperature," I said to Allen matter-of-factly. "We're at a balmy zero degrees."

The shuttle soon arrived, transporting us through a series of sleepy neighborhoods. A thin cloudy veil hovered twenty or thirty feet above the ground like a mysterious blanket. That marked an inversion, or a layer of acutely warmer air resting atop bone-chilling surface air. It trapped smoke, moisture, and pollution below.

I looked behind the van. Its exhaust stayed still, hanging in the air as if too cold and too tired to move. It formed a foggy trail in our wake that lingered for ten seconds or so before dissipating, reminiscent of an airplane contrail. The silence of the night was punctuated by the crunch of snow beneath the tires, as if the van was struggling to gain its footing.

Ten minutes later, it became apparent why. I stepped out of the van onto a four-inch-thick mound of ice caked to the roadway, involuntarily coming the closest I've ever been to performing a split. Ever the flexible gymnast, I winced in pain. Discarded cigarette butts littered the ground around the hotel's front stoop, each burrowed an inch deep in the ice where their slowly dying warmth had melted a depression. Chunks of rock salt provided some grip beneath my worn-out sneakers.

The dimly lit hotel room was small: *cozy* was probably the right word. A bowed-out cardstock print depicting an oversaturated northern lights display hung in a cheap picture frame roughly centered between two twin beds. A pair of neatly stacked pillows rested atop each. I grinned foolishly, beaming at Allen.

"Wasn't this such a bargain!?" I prodded. He rolled his eyes and smiled, slinging his backpack onto the bed nearest the door. I was always next to the window.

—

The next day started early, since we both worked East Coast hours. That meant an early morning wakeup at 4:45 A.M. It's hard to wake up when it's sub-zero, and I don't like waking up before the sun, either. This was a double whammy.

I set to work typing articles while Allen groggily opened his laptop and fired up PowerPoint. After thirty seconds of fumbling around and typing, he collapsed back onto his bed and retreated beneath the covers, wrapping the comforter around his head like a nun wears a habit. A blank slideshow was playing on loop, displaying TITLE TEXT HERE on a white background; it kept his laptop awake so he would appear online. He was getting paid for this (and more than me!). I scowled at him, but he couldn't see. He was already fast asleep.

By noon, I was wrapping up for the day, getting ready for a nap. Allen, well-rested and bright-eyed, emerged from his cocoon, energized from his hours-long slumber.

We ventured outside into the piercing sun, the backdrop of white snow proving blinding.

"Noodle House," Allen said with a grin, pointing at a neon sign dangling from a single-story home adjacent to the hotel. Its maroon-painted siding and yellow trim were peeling, weathered by decades of harsh Alaskan winters. I could sympathize, and I had only been in town eighteen hours.

Bing! echoed a doorbell as we entered. My eyes struggled to adjust to the darkness. Rows of vacant tables stretched along a tiled floor, each with a rubber tablecloth stapled to it and adorned with a rolled-up bundle of silverware. We sauntered over to one next to a space heater, its welcoming hum as inviting as the warmth it radiated.

"Hello!" a short Asian woman sang, scurrying to the table with a pair of menus. I smiled and asked her name. "Lanoi," she said, inquiring where we were from. Before long, we were deep in conversation. She explained that she had immigrated from Thailand and ran the restaurant herself, and that the pandemic had hindered business.

"How come you two are here?" she asked. I told her that we were looking for the northern lights.

"Amazing," she said, her eyes widening. "So colorful. I hope you see them." *Maybe the aurora wasn't just hype drummed up by the tourism industry after all,* I thought. *Even the locals seem impressed.* Allen and I set about hatching a plan to venture out in pursuit of them later that evening.

After a three-hour nap, we awoke shortly before 6:00 P.M.; the sun was already dipping below the distant hillside, bathing the valley below in a muted orange light. Fairbanks sits within the center of a topographic bowl, making forecasting local weather conditions a complex, and often Sisyphean, task. I gathered my camera gear, unplugging batteries and piling into a backpack while double-checking I had all my favorite lenses. Allen and I bundled up, then headed out in search of the lights.

It's about now I should probably remind you that most rental car agencies don't exactly make it easy for someone under twenty-five to procure a vehicle. Either you're forced to pay a jacked-up price and an obscene young renters surcharge or you're flat-out denied. Being notoriously cheap, I thought up an alternative: I rented a cargo van from U-Haul. My methods are rarely pretty or conventional, but they've almost always gotten the job done.

"All aboard!" I hollered to Allen, smirking as he climbed into the passenger seat. I tossed my camera in the cargo hold and then hopped into my seat, shivers immediately running up my spine. It then occurred to me that the van had been sitting out in minus-fourteen-degree Fahrenheit temperatures for the past twelve hours, and was in no hurry to heat up. We were riding in an icebox.

I clasped the icy steering wheel with my gloves, shifting into reverse. The two-wheel drive van struggled backward, groaning as the tires worked in vain to gain traction. I pressed the gas pedal more firmly. *Vroom!* We rocketed backward as if, after careful consideration, the van had decided to do its thing. "Weeeeeeeeee!" I remarked to Allen.

It was only 6:30 P.M., but the town appeared abandoned; traffic lights cycled through their rotation, supervising otherwise desolate roadways. Our goal was to escape city lights, which meant driving west along Alaska Highway 3. I was optimistic that the translucent deck of cloud cover would thin outside the city.

ABOVE: My nerves were high on the evening of June 24, 2013, the night before I flew to Nashville for my first presentation at an American Meteorological Society conference. The atmosphere offered me a show of goodwill, distracting me and assuaging my fears by delivering the most prolific lightning display in years to visit Cape Cod Bay. Upward of sixty cloud to ground lightning strikes slammed the water's surface every minute. My mother drove me to the beach as I fumbled with a twenty dollar Kodak point and shoot camera. I'd never even heard of a "long exposure" before, but by some miracle I captured this shot. From then on, I was hooked on photography. BELOW: February 21, 2014, brought unusually mild air to southern New England ahead of a sharp cold front that knocked temperatures from the 50s into the 30s. That sparked a line of strong thunderstorms with gusty winds and frequent lightning. The linear nature of the storms meant horizontally expansive electric fields and the potential for spider lightning. This lucky capture I snagged depicts lightning and remnant snow on the ground simultaneously.

ABOVE: My first "real" supercell thunderstorm—May 16, 2017 near Sayre, Oklahoma. It was isolated, discrete and spinning like a top. Dual channels of "inflow" appeared like arms as the twisting tower inhaled air in an inward spiral. Softball-sized hail was falling beneath the storm's base at the time of the photo; the lowered wall cloud, visible at the center of the storm, eventually touched down and produced a deadly rain-wrapped EF2 tornado in the town of Elk City. BELOW: A pyrocumulonimbus cloud—or a thunderstorm cloud spawned by heat released from a wildfire—looms over Elk City, Oklahoma on May 12, 2018. I had been in Moore, just south of Oklahoma City, but raced two hours west as soon as the smoke cloud began spitting out lightning strikes. I was treated to the most otherworldly display of pouch-like "mammatus" clouds I had ever witnessed, with pockets of sinking air hanging down from the dirty thunderstorm's anvil.

In tracking the fire-induced thunderstorm near Elk City, I was struck by the burnt brown hues wafting overhead in the rust-colored smoke plume. I attempted to take a picture of the sunset, but wound up with more than I bargained for when a close-range purple lightning strike photobombed prominently centered.

This lucky shot came on May 17, 2019 moments before the McCook, Nebraska tornado formed. The wall cloud, front and center, is obvious. A "clear slot" of bright skies devoid of overcast is visible punching into the circulation, marking where the rear flank downdraft was wrapping around to tighten rotation and squeeze out a tornado. The funnel at the bottom touched down less than two minutes later.

The Salar de Uyuni in Bolivia is the world's largest salt flat, topping the charts at almost 4,000 square miles. Its nearly flat topography and bright white surface is ideal for the calibration of satellites. Mirages are common in the desert environment, including this one I experienced, which appears to show floating mountains and vehicles. The thunderhead in the distance is hundreds of miles away over the Brazilian rainforests.

A wide-angle image depicting the textbook structure of the storm—the upended tornado being stretched to the northeast, the well- illuminated clear slot suffocating the funnel, and a wall of dust being kicked up by both the twister and the lagging rear flank downdraft.

ABOVE: The second tornado of May 17, 2019, near Stockville, Nebraska. A few splashes can be seen in the field a hundred or so feet in front of me. That's from hen egg–sized hail falling into the waterlogged grass. BELOW: A wall cloud descends from the base of a supercell thunderstorm shortly before producing a brief EF0 tornado near Okmulgee, Oklahoma, on May 22, 2019. I had just wrapped up a visit with Kelby when Michael and I rapidly blasted south to intercept the storm.

ABOVE: Dual boundaries set to clash on June 4, 2019, over Clovis, New Mexico. When the photo was taken, I was directly beneath a dryline—hence the clouds in the foreground. In the background, towering cumulus clouds anchor themselves along an arriving cold front. BELOW: The two boundaries in Clovis ultimately collided, producing a band of severe thunderstorms that dropped torrential downpours and emitted incessant pinpoint lightning strikes over the open chaparral.

ABOVE: An out-of-bounds dust storm approaches Lubbock, Texas, on June 5, 2019. Dust storms, more properly called *haboobs*, are generally relegated to the Desert Southwest. BELOW: A photograph of the July 2, 2019, total solar eclipse in the rural mountains northeast of Vicuña, Chile. We removed our protective glasses seconds later. The sun's corona, or atmosphere, is visible radiating outwards from behind the moon's shadowy occultation. The last pinpricks of direct sunlight, known as Bailey's beads, have congealed into a "diamond ring" effect.

ABOVE: A broader view of the total solar eclipse over the nearby mountains. The sun was only 11 degrees above the horizon, making framing the eclipse against an artistic terrestrial backdrop more easy. The curved shadow of the moon can be seen centered over head. It was approximately 90 miles in diameter. BELOW: Sixty mph winds sweep through a field in southwest Oklahoma on May 13, 2020, amid a severe thunderstorm producing hail. The fiery crimson sunset protruded from beneath the cloud base, making for an auspiciously timed moment of peace.

A classic mothership supercell looms near Burkburnett, Texas, on May 22, 2020. At the time, I was standing on a twenty-foot pile of sand praying I wouldn't be struck by lightning. Apparently the prayers worked, because I'm still around and have this wicked shot to prove it. The funnel cloud on the bottom right next to the high voltage power line eventually birthed a brief EF0 tornado.

ABOVE: A mystical blanket of "hail fog" southeast of Rapid City, South Dakota, on June 4, 2020. A second approaching supercell thunderstorm with large hail is visible in the distance stalking its predecessor. BELOW: The northern lights dancing through the skies over central Alaska as seen from Alaska Airlines flight 234 on March 20, 2021. Allen had already arrived in Seattle via the previous flight, but I was skimming along the underside of heaven at 35,000 feet.

ABOVE: Light pillars that Allen and I encountered upon returning to Fairbanks after taking a dip in the Chena Hot Springs 62 miles to the east. CENTER RIGHT: I work to affix the hail cage to Ridgie, my Honda Ridgeline, on April 23, 2021, prior to convective initiation. BELOW: A wall cloud looms near Lockett, Texas, shortly before producing a tornado on April 23, 2021.

ABOVE: A stovepipe EF2 tornado swirls through fields near Lockett, Texas, on April 23, 2021. We were tardy—arriving from the west as the tornado breezed across the road and corn husks rained down from the sky—but it's better late than never. BELOW: The second tornado near Lockett, Texas, on April 23, 2021. The dusty elephant-drunk funnel was snorting red dirt into the air. Tennis ball–sized hail pelted us moments after I snapped this picture.

ABOVE: Lighting crawls along the underside of a sunlit thunderstorm on April 23, 2021, following a siege of violent weather near the Red River of Oklahoma. BELOW: A classic supercell thunderstorm near Sterling City, Texas, on May 17, 2021. We were peering west toward the rotating updraft, beneath which no rain or hail was falling at the time. A battering of baseball-sized hail eventually swirled around the tornado.

ABOVE: Dry air near the surface meant that, though the tornado had touched down, a "condensation funnel" wasn't visible all the way to the ground. Here the tornado can be seen menacing a wind farm. BELOW: A tornado materializes near Selden, Kansas, on May 24, 2021 on what, for me, was an unexpected last- minute storm chase just a few minutes from my hotel. A collar of eerie aquamarine wraps around the pendant funnel, dressed in a scarf of scuddy gray.

A multiple-vortex EF1 tornado lunges across the road in Selden, Kansas, on May 24, 2021. Notice the individual funnels all orbiting a common center. Each locally enhances wind speed and results in narrow strips of more severe damage.

SUBWAY

An incredible mammatus-filled sunset following the destructive storms of May 24, 2021, in Colby, Kansas. Kelby and I were at El Dos de Oros Mexican restaurant when the light indoors suddenly changed as if the building was on fire. We rushed outside to witness departing storms to the east bathed in a delicate sunlit hue. Pockets of mammatus could be seen sagging beneath the anvil.

ABOVE: A surprise supercell thunderstorm near Colby, Kansas, around 10 A.M. on May 26, 2021. Powerful morning thunderstorms are extremely rare. This one dropped tennis ball–sized hail on a morning that was advertised to be mostly sunny. BELOW: A "low precipitation" supercell—or a rotating storm stripped of its rain—spinning over open fields north of Colby, Kansas, on May 26, 2021. The storm, which went unwarned by the National Weather Service for more than a half hour, dropped lime-sized hail east of State Route 28 over open fields.

The low precipitation supercell passes overhead near Atwood, Kansas.

ABOVE: A leap of faith—skydiving from 15,000 feet above Titusville, Florida, on June 18, 2021. Allen took the plunge shortly after I did. BELOW: My introduction to TV broadcasting—live tornado coverage on FOX 5 DC on the night of July 1, 2021.

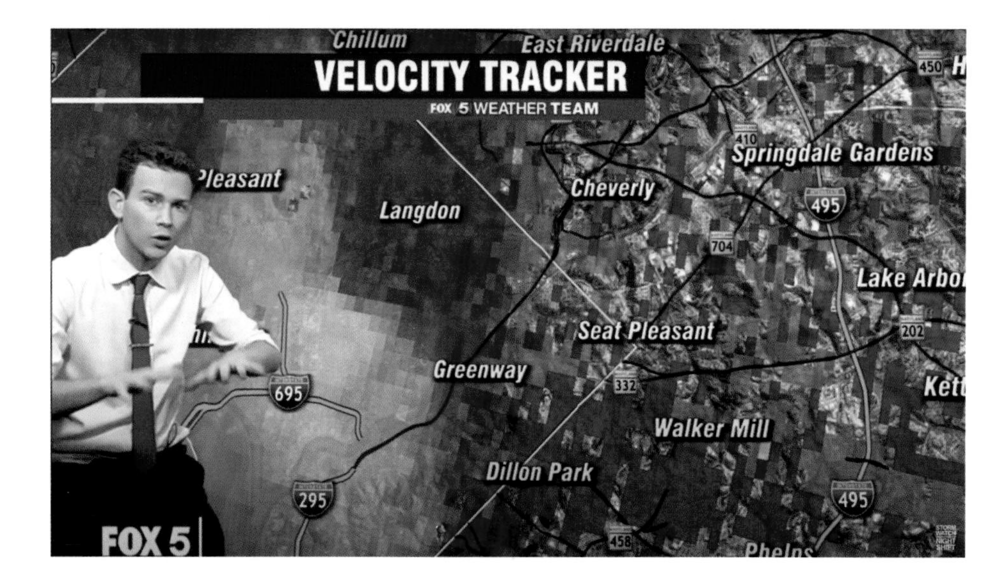

TOP RIGHT: Me delivering my first full on-air forecast on July 3, 2021. CENTER: An EF2 tornado causes damage over Weems Creek near Annapolis, Maryland, on September 1, 2021. I was driving along Rowe Boulevard at the time. BELOW: Me happily at the helm at FOX 5 DC on December 28, 2021.

ABOVE: Allen and I stand beneath a shooting star in the Atacama Desert of Chile during the 2021 Geminid meteor shower. The Milky Way can be seen to our left. BELOW: An areal shot of the Perito Moreno glacier along the Chilean-Argentinian border on December 16, 2021.

We drove for about twenty minutes, bantering back and forth like deranged, overtired news anchors. A sign reading ESTER GAS shone in elegant white letters, illuminating a chipped patch of parking lot on the edge of the road. Fuel prices were displayed in green and red. I turned right onto Old Nenana Highway, intentionally speeding up halfway around the curve to make the back tires swing outward in a controlled skid.

"Weeeeeeeeeeee," I jeered again, prodding Allen. He smiled and stifled laughter, passively along for the ride.

The convenience store's lights became dimmer in the rearview mirror as we climbed a hill. A side street, Gold Lode Road, looked promising. The roadway became narrower as the mounds of snow towered higher. We climbed steeper and steeper, the van, which Allen nicknamed "Fridgie," moaning in protest. I pushed the pedal to the floor, to no avail. The tires were squealing but we were making little progress. Eventually, we slowed to a crawl before creeping backward.

"I guess this'll do," I laughed as we slid down the hill, Allen appearing startled as he peered out the window. I shifted into reverse and cranked the wheel to the right, briefly tapping the gas to spin the van around. Spying a gravel driveway at the berm of the slick road, I limped the van to the left and set the emergency brake.

"Land ho!" I hollered.

Excitedly, I killed the ignition and turned off the lights, hoping to see the aurora appear overhead. Instead, utter darkness greeted us, a few pinprick stars twinkling in the celestial abyss.

I slipped outside for a closer look, the cold taking my breath away. Within minutes, a painful numbness stung my extremities before expanding from my fingertips through my hand, chilling my wrist and forearm. I abandoned my ill-advised optimism for stark realism, realizing I urgently needed to get inside the van. I fumbled at the door for fifteen seconds before managing to latch my dormant fingers around the handle.

We sat there for an hour, huddling against the strained heating vents like moths swarming around a lamp. I was about ready to give up when Allen pointed to the sky. "Is that something?" he asked?

Indeed it was. A pale brushstroke of green shimmered overhead like an eraser smudge, initially indistinguishable from cloud cover. As it intensified in brightness, a second, more horizontally extensive band began to climb over the horizon. It wasn't much, but we were in business.

Long before the northern lights ever graced postcards or elicited awe-struck shouts and screams from sky watchers, their story begins ninety-three million miles away on the surface of the sun. The northern and southern lights, or aurora borealis and aurora australis, respectively, are a bona fide form of space weather, and require a working knowledge of solar physics to forecast.

Think of the sun as a high-intensity lava lamp. In the lamp, bubbles are heated at the bottom, becoming less dense and rising to the top before eventually cooling off and subsiding. The sun is sort of like that. While the surface is eleven thousand degrees Fahrenheit, its core is a giant self-contained nuclear fusion reactor where temperatures reach tens of millions of degrees.

Bands of magnetism wrapping around the sun in horizontal latitudinal belts interact to form pouch-like regions of more concentrated magnetic field. These pockets ascend in the outermost layer of the sun, reaching the surface like bubbles in a pot of boiling water.

When they do, sunspots result. The cool, bruise-like discolorations are visible from Earth and crackle with magnetic energy. Some of the energy pours out into space, while other sunspots throw out magnetic field that reconnects to the sun in loop-like structures. The magnetism itself is invisible, but is often wrapped in solar plasma, making it resolvable by space-borne satellites.

Once in a while, interfering magnetic fields interact explosively, hurling a burst of energy into space. Like the pent-up energy contained in a stretched rubber band, the contorted magnetic fields realign and reorient in spectacular fashion.

That first produces a brilliant flash of light known as a solar flare, which ejects high-energy particles into space that can reach Earth in just a few tens of minutes. They travel at almost the speed of light, and can disrupt electrical and GPS signals and spur temporary shortwave radio blackouts. Their effects are isolated to the sun-facing side of Earth, but can wreak havoc on navigation and aviation. By the time we know about one, it's often too late to do anything about it.

Each energetic release is also accompanied by a more targeted coronal mass ejection, during which magnetism, material from the sun's atmosphere, and other particles surf an interstellar shockwave of sorts through space. It takes two to four days to reach Earth, but can have dramatic effects as it interacts with our planet's own magnetic field, or magnetosphere.

At the Earth's core is an enormous 750-mile-wide ball of iron and nickel. It's red hot, and would be molten liquid if it wasn't for the extreme pressure squeezing it into a solid. Its edges are gooey and viscous, however, allowing it to spin. That rotation spurs a magnetic field, making the Earth into its own large electromagnet.

When energy from a coronal mass ejection (CME) reaches Earth, it reverberates off the Earth's magnetic field. Like any magnet, Earth has a north and south magnetic pole, roughly collocated with the geographic north and south poles.

Our magnetic field is most intense around the poles, since that's where lines of magnetic field converge. The magnetosphere is like Earth's natural sunscreen. Without it, we'd be cooked by high-intensity rays and dangerous solar energy. Instead, the energy from CMEs is transformed into harmless visible light. More intense geomagnetic storms overwhelm the poles with energy, forcing the northern lights down to the mid-latitudes.

The Space Weather Prediction Center in Boulder, Colorado, is staffed 24/7 with forecasters who monitor the sun's every pop and crackle. While sunspots are most common every eleven years at the peak of the solar cycle, any sunspot, even during a nearly spotless stretch of tranquil activity, can erupt with little warning. We were heading to Alaska at solar minimum, the next peak not scheduled to arrive until mid-2025, but I bet on the sky showing off anyway.

For several nights, the lights offered only a tease. I was snapping some great photos, but the displays were underwhelming in real life. I figured this was as good as it would get.

The third night fell on March 14. Pi Day. To most people, 3/14 is just another mundane late winter day, but for math and STEM nerds, the date's digits are cause to celebrate. While en route to our typical aurora-watching spot, I suddenly drove into the parking lot of a McDonald's on the edge of town. Allen turned and looked at me, his brow furrowed.

"I thought you said you weren't having any more French fries this trip," he said, referring to the half pound of curly fries I had devoured the previous evening. I had sworn off any fries for the remainder of the week.

"It's Pi Day. I'm getting us pie," I said gleefully, making an exaggerated smile as we pulled up to the drive through speaker in our creaky cargo van. I recited our order like a child proudly picking out his own birthday gift.

"You're ridiculous," Allen laughed, rolling his eyes and smiling. For $2.37, we were going to celebrate Pi Day the right way.

Once again, the sky failed to deliver, and, two hours and a pair of cardboard box of mini pies later, we dismally began the quiet ride back to the hotel. Then five minutes before our arrival, the sky exploded in color.

"Wait, Allen, look up!" I shouted, even though he was sitting barely three feet away. I hastily tapped him on the shoulder in case he didn't hear me.

"Woah," he marveled, looking up at a ribbon of pink bright enough to slice through the light pollution from streetlamps, traffic lights, and businesses on the western edge of Fairbanks. It slowly undulated, like an ocean wave made out of molasses. I focused on the roadway while Allen craned his neck for a better view.

"We've got to get back to Ester!" I exclaimed, searching for a spot to make a U-turn. Seconds later, we were rocketing down the desolate state highway back toward our secret spot. The chase was on.

My eyes were glued to road, but that didn't stop me from stealing periodic glances at the sky. Besides, it was bright enough that, even focusing on the roadway, the glow of the aurora was visible in my periphery. I caught a glimpse of a lime green swirl of light rippling above, pillars of light dancing around one another and tracing out the shape of a cinnamon bun. The tops were frosted with a soothing pink tone.

Gorgeous colors enveloped the sky, albeit briefly. By the time we made it back to our post, the fleeting Technicolor outburst was simmering. I figured that was it, the sudden geomagnetic perturbation beginning to relax. It reminded me of a swimming pool. Even if you cannonball into one, the waves eventually settle and the pool goes back to normal.

With my camera finally mounted on its tripod and a wide-angle lens attached, the display fell flat.

"Do it again! I wasn't looking!" I yelled at the dull night sky, my words of frustration disappearing into the frosty Arctic air as my condensed breath disintegrated. Allen remained in the van. I trudged through the snow, aggressively tossing my tripod in the cargo hold. Summiting my way into the elevated cab, I climbed into my seat. I stared at the sky longingly until Allen finally broke the silence.

"At least you got to see them, even if you didn't get any pictures," he said softly. "That's something."

It was now 2:26 A.M. I replayed the evening in my mind, thinking through anything I could have done differently. It occurred to me how truly strange of a trip I was on. I was with a close friend and adventure buddy whom I had met on the internet (and who actually laughed at my jokes), now roaming the Alaskan tundra in a U-Haul van, it was the middle of the night and minus-twelve degrees Fahrenheit outside, and we had just feasted on miniature pies beneath a mesmerizing display of the northern lights. The rest of the world was gripped in the midst of a pandemic, but I was living in a separate universe, one of wonder and peace.

"I guess there are just some things in life meant to be lived rather than photographed," I said, more for myself than anyone else. Allen nodded before I continued. "These are going to be some memories."

Making Memories

Making memories is something that has always fascinated me. I've learned that one can't actively go out and seek to make memories—they just happen. The most special moments in life are those that can't be planned, anticipated, or replicated. Like the weather, sometimes things come together just right; those are the moments that give life its meaning.

I have family members who spend thousands of dollars, flocking to all-inclusive resorts and building their days around a scheduled set of activities they think will create memories. Sometimes it works, but that's never been my style. I don't need a special activity to make a lifelong memory; that would feel forced. To me, life is about surrounding yourself with the people who make ordinary everyday moments extraordinary.

My cobbled-together trip to Alaska featured a lot of those moments. With a good adventure buddy, even the mundane quickly became jam-packed with hilarity.

With little to do and feeling bored on the fourth night, for instance, Allen and I decided to head to Walmart. That ended in the two of us giggling hysterically as we performed doughnuts in the van on the ice-coated vacant parking lot. Even though it was midnight, we decided to buy a cake from the bakery for good measure, then devour it in the parking lot of a bingo hall.

It turned out that Chena Bingo was open late, with a session at 11:45 P.M. Allen didn't need any convincing.

"I'm down," he said with a grin. I affixed the plastic cover onto the cake tin, wiped the frosting off my face, checked my reflection in the side-view mirror, and hopped down from my perch in the van. We sauntered inside like two normal young adults heading into a bar for an evening on the town. Instead, we were spending our Friday night at a bingo hall, and I wouldn't have it any other way.

Chena Bingo smelled like a bowling alley; the familiar scent of aged carpet, fried food, and permanent marker greeted us as we entered the aging lobby. I looked around.

"I think we're the only people here not in AARP," I whispered to Allen.

"Doesn't that mean you'll fit in perfectly then?" he joked. I laughed, partly because it was funny and partly because it was true.

We bought a couple ink daubers, each paid $10 for five rounds of play, and settled into our seats. Before long, the numbers began flying: I23, O63, G47, B8 . . . we were struggling to keep up. We had five bingo sheets between us total, each of which had half a dozen individual game cards per round. That meant thirty bingo cards to check for every single number.

"Let's share one," Allen said, placing a sheet of bingo cards in between us. At first, it seemed like a good idea—whoever finished his individual cards first took care of daubing the communal card. Of course, our childish natures soon took over, and what started as an exercise in efficiency became a bingo battle to the death.

Before long, the communal bingo card was tie-dyed with a minefield of errant daubs; blue ink stretched down Allen's forearm, while green dots lined my wrists where he nabbed me with his dauber instead by mistake.

"So you have aim issues," I teased.

Suddenly, I noticed something remarkable: I had five in a row. *Bingo.*

"Look at this!" I whisper-shouted to Allen, gesturing toward my pockmarked bingo card. His eyes rested on the five diagonally connected ink splotches sprawled across the sheet.

"Double check those numbers," he said with a smirk. "You got a bit violent."

"I think this one is the $1,000 game," I said to Allen, cross-referencing my numbers against those displayed on a large ticker board suspended from the ceiling. Sure enough, I had it.

"Bingo!" I screamed without hesitation, abruptly catapulting out of my chair with such force that it toppled to the ground. I thrust my card into the air in one hand, gripping my royal blue dauber in the other; I looked like a bargain-brand Statue of Liberty. Silence engulfed the room.

Every eye in attendance turned to me. A series of rhythmic clanks echoed out as players slammed their daubers down in frustration. It didn't matter—*I* was going home with $1,000. A muffled static emanated from the interrupted bingo caller's microphone. I was bouncing on the soles of my shoes, an ear-to-ear grin plastered on my face.

"False Bingo!" a raspy voice jeered in the back left corner of the hall. My smile vanished. The hall erupted into alternating shouts of "false bingo!" and "players, hold your cards!," each accusation more hostile than the last.

I yanked my arms out of the air, dangling them sheepishly at my side. The middle-aged man who had been calling numbers at the front adjusted his wire-rim spectacles, sighed, and began rattling off numbers again. I turned to Allen.

"Did you double-check those numbers like I said?" he asked, his lip quivering as he struggled not to laugh. I felt as if I had been hit by a truck.

"I don't get it," I croaked softly, like a kindergartener who had dropped his ice cream cone. *But I got five in a row!*

"There haven't been enough *I* numbers called," a woman behind me affirmed. I turned. She appeared to be about seventy years old, but wasn't frail. In fact, she looked tough, like a no-nonsense librarian who was about to reprimand me.

I handed her my card to inspect, but before I could open my mouth to speak, she pushed it away.

"This ain't the bingo you're thinking of," she said. "We're doing a four-corner postage stamp."

She laboriously pointed to the large display board, her outstretched arms adorned with metal bracelets and wristbands. They clanked against one another like a venerable wind chime.

"You need four in each corner," she concluded gruffly before shifting her gaze back to her own cards and uncapping her dauber. I shrank back in my seat. Allen was cackling.

"Any minute they'll show up with torches and pitchforks," I mumbled, clenching my jaw while simultaneously choking back laughter. "My credibility is shot."

The following days proved equally laughter-filled. We went snowmobiling, toured a public park (or what we assumed to be one, but it was buried in snow), and nearly got kicked out of an art and culture museum (I was imitating the voices I associated with each portrait, and Allen couldn't stifle his shrieks of laughter). I was having such a good time that I forgot to be disappointed in the unshakeable cloud cover preventing another glimpse of the lights.

On our final full day in Fairbanks, we decided to visit the Chena Hot Springs, located about an hour and fifteen minutes east of town. To say it was located in the middle of nowhere would be an understatement. The fifty-mile-long paved highway was called Chena Hot Springs Road, and, after a few minutes of driving, it occurred to me we hadn't seen a single other vehicle.

The sun was low in the midafternoon sky, periodically dipping behind the peaks of mountains. Elongated shadows stretched across the forested landscape, snow-studded pine trees standing on the edge of the roadway like patient hitchhikers. We crossed over a river about a hundred feet wide, presumably supplied by daytime snowmelt. Its rocky banks extended out far on either side of the water; I imagined that, in just a few weeks as the warm season arrived, it would be on the verge of overflowing.

The road got narrower, the pavement replaced by loose gravel. We were getting close. We passed beneath a wooden archway that read WELCOME TO CHENA HOT SPRINGS RESORT, easing to a stop next to a full-size RV. Grabbing our bags, we began strutting toward the resort's main lodge. I locked Fridgie.

The hot springs were like a sauna in the middle of a skating rink. They were surrounded by a wall of boulders. Steam poured out of the springs into the dry Arctic air, instantaneously forming a rime ice on the rocks that accreted several inches thick. I felt like I was in a Scandinavian fortress.

Time passed slowly in the hot springs, the bath-like warmth melting away any subliminal cares, worries, or tension. I hardly noticed when the azure blue curtain of twilight descended on the darkened landscape.

"We should go," I said.

Steeling ourselves against the cold, we exited the welcoming waters and tiptoed through the piercing cold standing between us and the lodge. My hair froze during the thirty-second speed walk, my damp bathing suit clinging to me like a searing massage of liquid nitrogen. After changing and donning our winter gear once more, we trudged to the restaurant, splitting a trio of kids' meals before venturing outside to return to Fridgie.

"Ope, there it is!" I announced, staring at an arc of luminance above. It was only 9:00 P.M., and plenty of night remained. We hastened to the van, fired it up, and began the drive back to Fairbanks, searching for an open field or turnoff from which to enjoy the nascent show.

We stayed until a cloud swept overhead, thick enough to blot out the aurora but thin enough that the moon cast a ghostly corona through it. Snowflakes glided down to the ground like a peaceful confetti.

Satisfied that we had seen everything we were going to, we began the long drive back to Fairbanks; less than two minutes up the road, the cloud vanished, and stars once again appeared. The snow was nowhere to be found.

The forecast had called for a 20 percent chance of isolated snow showers. Apparently, they were very low-topped, the moisture that gave rise to them trapped within 2,000 feet of the surface. Little did I know how much of a role that same moisture would play later on closer to Fairbanks.

We drove on through the dark, my inner child treating the empty winding roadway like rainbow road in a game of *Mario Kart*.

"Thanks for a fun trip," I said to Allen. I meant it—I had never laughed so much in a single week.

"Was there anything you didn't get to see or do up here that you wanted to?" I asked. He looked down at the dashboard and thought for a moment, then grinned.

"A moose!" he said, then laughed. All week long he had been talking about his desire to see a moose.

"Ask and you shall receive," I said matter-of-factly. "You want a moose, I'll get you a moose. Give me one hour."

"Oh jeez!" I shouted, sitting bolt upright in my seat and gripping the steering wheel with two hands. I quickly jerked the wheel to the left to take the curve as wide as possible, eased my foot off the gas, tamped the pedal down rapidly, and swerved back into the right lane, making sure to avoid the patch of ice coating the double yellow line.

I tapped the brake as we ground to a halt, looked up, caught my breath, whispered "thank you," and chuckled at the universe's sense of humor. Only forty-three minutes had elapsed; the moose had shown up early.

"You good?" Allen asked, seemingly confused.

"What do you mean?" I asked. He had been staring out the window the entire time.

"Why did you swerve?" he asked. I was dumbfounded.

"Because your moose was in the road, you dodo!" I shouted, looking around as if for a hidden camera. My life was a sitcom.

"Wait, really?" he asked. I could tell he was being serious.

"Allen, we missed it by hardly eighteen inches. We are *very* lucky," I said. It had been smack dab in the middle of the right lane on an obscured bend in the road. The creature's antler nearly clipped our radio antenna; I estimated it at nearly one thousand pounds. Allen's moose had come within three feet of him, and, somehow, he didn't even notice. I wasn't letting him fly back home without seeing a moose.

Amused and exasperated, I shifted into reverse, making sure there were no headlights anywhere in the vicinity. We hadn't seen another vehicle for twenty minutes. I backed the van up about three hundred feet until I saw a hulking figure in the rear view mirror illuminated by the brake lights. It lumber across the road as I eased the van off the highway.

"Allen, this is Mr. Moose," I said, feigning introductions while I opened Allen's window. "Mr. Moose, Allen."

"Oh *wow*," Allen said, face to face with the moose he had requested. The universe had delivered. As opposed to that bad stretch a few months ago, I was on the karmic upswing again with the animal kingdom. I like to think helping out the lizards and turtles was partly responsible.

I realized why he hadn't seen the moose: its eyes didn't reflect light, like the proverbial deer in the headlights. Instead, his eyes were black, blending in with the somnolent nocturnal landscape and the dark pavement. The moose lazily waltzed off the road, seemingly uninterested in its new pair of bewildered onlookers.

"I'll get you a picture of your moose," I said to Allen. Opening the door and walking around to the passenger side of the van. My camera equipment was in the back.

"Don't get charged!" he shouted, but I knew I was safe. The universe wouldn't send us a malicious moose, right? I opened the sliding door to the cargo hold, assembled a camera, ensured I was in NO FLASH mode, and began snapping photos. I figured if the moose got too close, I'd just shut the door and remain safely in the back of the van.

Instead, the moose stood still and tilted his head at an angle, posing for the camera. He was only twenty feet away. After a few photos, I figured it was time we bid the moose farewell; it was getting late.

"Bye, Mr. Moose!" I shouted, wandering back to the other side of the van and hopping into the driver's seat.

"Bye, Moose," Allen said softly. He stared out the window longingly, the moose receding into the background and once again disappearing into the darkness.

We crested the final hill leading into Fairbanks. It was almost 1:00 A.M. The sleepy city sat nestled comfortably in between mountains, any icy fog obscuring the city. Columns of light shone through the mist.

Wait a second. *Columns of light!?*

"Light pillars!" I hollered boisterously, inadvertently jarring Allen. He leapt up in his seat, probably afraid another moose was on its way to crash through our windshield.

"What's a light pillar?" he asked. That was my cue; I pulled the van to the side of the road. I could hardly contain my excitement.

"Hop out real quick," I said. Allen, as patient as he was adventurous, did so. I walked over and joined him next to the guardrail. The air was glistening.

"See these specks in the air?" I asked. He nodded. "Diamond dust. The air is so cold that what little moisture there is in it deposits as ice crystals. It usually happens at temperatures below ten degrees."

I went to pull my phone out of my pocket, but Allen was already a step ahead of me.

"Portrait or landscape?" he asked knowingly.

"Landscape," I replied, internally grateful I had found a friend who knew me so well, and was tolerant enough to put up with my shenanigans. "Thank you."

He nodded, letting me know he'd tapped RECORD. I launched into my enthusiastic explanation of dew point and the capacity of air to hold moisture and ice crystals, my hands waving around like the dancing tube man inflatable usually found outside used car dealerships. I couldn't help myself.

"The ice crystals are all shaped like hexagonal plates," I stated to the iPhone, gesticulating around me as if I was pointing to a weather map. "They act like little mirrors. That means light shining upwards from a source gets beamed back down toward us. The result is a column of light."

The air was clean, which meant that the moisture in it existed as supercooled water droplets, or liquid droplets present at subfreezing temperatures. Water will only freeze once it has something to freeze onto. In extreme circumstances, supercooled water can remain a liquid at readings as low as minus-fifty-five degrees Fahrenheit. But at ground level, the liquid droplets easily crystallized to deposit an icy glaze on roads, sidewalks and any untreated surfaces. Meteorologists refer to it as freezing fog.

We were standing on the shoulder of the Steese Highway next to a U-Haul storage facility. Bright white lights spaced thirty feet apart hung along each of the seven storage buildings. Each had a narrow pillar of luminance hovering over it. I knew the actual pillars weren't really occurring over the lights, but rather only appeared to be there. The individual ice crystals responsible were actually located about halfway between each light source and ourselves.

A new yellowish-white column appeared half a mile down the road, splitting into two. The pillars were moving, their origins tracing a path down to below a thicket of trees. Eventually, a car emerged from behind a bend in the roadway. The pillars became brighter as the car approached, suddenly dissipating as it passed us about fifty feet away.

I snapped dozens of photos, finally tiring as the cold became inescapable. After four or five pit stops on the way home, it was time to get back to the hotel. Of course, that didn't prevent me from kidnapping Allen and driving all around town to gawk at the light pillars the same way I oohed and aahed at Christmas lights as a child. Allen, indefatigably patient, sat to my right in a vegetative state somewhere between sleep and hunger.

"Ooh, we have those Costco muffins back at the hotel we need to finish!" I said eagerly. Allen shook his head and smiled. At 2:20 A.M., we pulled into the parking lot of the Bridgewater Hotel. I grabbed my camera and we marched inside, still marveling at the light pillars all around.

"What the hell?" Allen said as we entered the dark elevator, his gaze resting on a small poster on the wall advertising the Fairbanks Bureau of Tourism. It showed a pod of narwhals, but his face was Photoshopped onto of their bodies. He burst out into laughter. I had forgotten I'd made that earlier before we left.

"You're ridiculous," he said, shaking his head and laughing.

—∾—

The next day, Friday, began with brilliant sunshine. We enjoyed a last trip to Noodle House, said good-bye to Lanoi, and checked out of the hotel. I couldn't help but feel gloomy as I marched to the U-Haul to drive Allen to the airport. I never have done well with returning home from an adventure, but I try to look forward to the next one.

"Storm chasing is only a month away," Allen said, smiling reassuringly. I spend all year long looking forward to my month on the Great Plains; this year, I'd have a copilot as madcap as I.

We had some time to kill before Allen's flight. He was slated to leave for Seattle at 3:25 P.M. and I didn't take off until 1:30 A.M., so we decided to swing by Petco on the way to the airport. We browsed the fish, perused the lizards, and tried to get the parrots to talk. No dice.

Finally, it was time to go. We hopped in Fridgie one last time, only to find that the power steering had somehow given out. We could either call for a tow truck, wait for Fairbanks's only Uber, or try to drive and wing it. I called my father, the automobile guru, and asked if it was safe. Sure enough, it'd be fine for a short journey.

"Meh, it's only six miles," I said to Allen, turning the ignition. With great effort, I tugged the wheel to the right. The tires turned slightly.

"Fridgie doesn't want you to go," I said to Allen, smiling.

"I don't want to, either," he said. We were both scheduled to return to Washington, DC, on the same connecting flight early in the morning; the only difference was that Allen would have a twelve-hour layover in Seattle and enjoy a good night's sleep. (I knew he would stubbornly insist on staying in the airport, so I reserved him a hotel room.)

After dropping off Allen, I drove silently to the U-Haul terminal off College Road, parked in between two box trucks, and collected my belongings from the cupholders and center console. I lugged my baggage into the rental center.

"So, did you see them?" a familiar voice greeted me. It was Aaron, arguably the friendliest store clerk I have ever met. He was the first person

I had interacted with when I arrived in Fairbanks a week prior; a basic rental pickup turned into a half-hour conversation, an exchange of social media information, and an invitation for Allen and I to dine with him and his wife. He even offered his spare guest rooms for the next time I was back in town. Southern hospitality had nothing on Alaska.

"Kind of, but they weren't that good. Are they always like that?" I asked.

"Oh, then you haven't seen a real display," he said. "You know one when you see one."

It was now only 4:00 P.M., and I had nothing to do until midnight, when I had to be back at the airport. I took an Uber back to a restaurant next to the hotel I had checked out of and scarfed down a mediocre pizza. I figured I might as well treat myself to a glass of white wine and a piece of chocolate cake; I was on vacation for another few hours.

I have a love-hate relationship with dining alone. On the one hand, I love the peace and quiet (and the fact that no one judges how much food I order), but I also get lonesome after twenty minutes of not talking to anyone. That's usually when I default to scrolling through my phone.

I pulled up SpaceWeatherLive.com, a dashboard that plots various metrics of space weather and the state of the Earth's magnetic field. The numbers immediately caught my eye—the interplanetary magnetic field (IMF) was strong, the BZ was negative, and the Kp index was climbing; it was already at a four! To the average person, the numbers may sound like mumbo jumbo, but they were enough to make my pulse quicken. Maybe I would see the northern lights after all, from 35,000 feet.

The Kp index is a number that corresponds to how perturbed the Earth's magnetic field is. It ranges from one to nine. The aurora can make an appearance in Fairbanks with Kp values of two or three; a four usually corresponds to a good display. Models were suggesting the Kp would climb to a five by midnight, at precisely the time I was taking off. The northern lights could spill all the way down to the US-Canada border.

That's when I realized the big issue: I didn't have a seat assignment. My only shot at seeing the aurora would come if I had a north-facing window seat on the left side of the plane. The odds of that happening on a fully booked flight were slim to none.

With another six hours to go, I went to the only place I could think of: Chena Bingo. That meant dragging sixty pounds of baggage, including a three-wheeled decrepit suitcase, a half mile. I alternated hands as I tugged the suitcase through pothole-like depressions in the ice and snow, arriving to Bingo a half hour later both sweating and shivering. Once again, I didn't win.

"You're 16A," the friendly Alaska Airlines check-in worker announced. She appeared to be in her lower sixties, with curly white hair, bright red lipstick, and the smile of a grandmother about to whip out a tray of cookies. Instead, she handed me my boarding pass and claim ticket for my checked luggage. "That's the emergency exit row."

She didn't know it at the time, but she was giving me a once in a lifetime experience.

I hastened through security, speed walking to my gate. I still had plenty of time before takeoff, but I felt that by walking quickly, time would move faster and I'd be in the air sooner.

Apparently, it didn't work that way, because we were slapped with a forty-five-minute delay. I groaned, watching the Kp index tick up to a five. I wanted to be in the air *now*.

Instead, I strolled over to the gift shop, knowing exactly what I was looking for: a stuffed moose. Allen's birthday was only a week away. I quickly located a six-inch-tall moose with big yellow eyes; FAIRBANKS was sewn onto his hoof. He'd love it.

At the gate, I made friends with a family of five from Arizona; they too had made a last minute trip to see the northern lights, but hadn't seen much more than the unremarkable displays Allen and I had encountered. I told them about the spurt of swiftly moving solar wind about to slam Earth's magnetic field.

"They'll be out there," I affirmed. I checked my phone: the Kp index had climbed to a six. The Earth was in the crosshairs of a full-fledged geomagnetic storm.

Finally, the time came to board. I slung my camera bag over my shoulder and donned my neck pillow around my collar. Passengers shuffled aboard, virtually everyone yawning and visibly drowsy. The flight attendants expedited their safety announcements, dimmed the cabin lights, and fastened their seatbelts. Unceremoniously, we rumbled down the runway, lifting off into the night.

I pressed my face against the Plexiglas window pane, alert for any possible hints of color. Where *were* they!? It occurred to me that the geomagnetic storm may be so severe that the auroral oval was actually shunted farther south. Then, something amazing happened.

The sky turned green. Not just a little green, but fully green. The postcards hadn't been lying after all! The pilot turned off the winglet lights, allowing for a better view. It was like he heard my subliminal pleas. Most passengers were asleep, and I was sure I was the only one looking out the window. I was shocked; these people were missing the sight of a lifetime!

That was just the beginning. I watched as the aurora transformed before my eyes, shifting shapes in a crescendo of elegance and caprice. I was transfixed; this couldn't possibly be real!

The diffuse but vibrant colors consolidated into a single band of interwoven columns, waves of pulsating luminance rippling through them like the fluttering of laundry strung out in the wind. Occasional eddies of glowing plasma swirled through the sky, the display unlike any of the comparatively meager shows I had beheld. Vibrant shades of purple, pink, and turquoise accompanied the green. I scanned the skies afraid to blink—I didn't want to miss anything.

It occurred to me that I had to get pictures—no one would believe what I was seeing! I immediately snatched my wide-angle lens and the body of my Nikon D3200, pressed it against the glass, and began clicking.

I quickly noticed the camera was picking up the reflection of small overhead lights on the no-smoking signs located in each row of seating. I placed my U-shaped travel pillow around the lens and tossed my jacket over my head, using it as a cloak to shield the window. The next picture would become one of my all-time favorites.

I was giddy. The amorphous light was dancing, a whirlpool of color swirling directly overhead. It was as if the atmosphere was painting with watercolors at time lapse speed. The symphonic display reminded me of an orchestra, with overtones and more diminutive acts overlapping in a harmonic frenzy. The fluid seemed somehow sentient, like the universe was alive; I was witnessing its synesthetic whisper.

It was everything I had hoped for and more, but amid my euphoria was a lace of guilt. Something was missing: Allen. He wasn't there to share in the moment. It seemed almost unfair that he was fast asleep in Seattle while I was experiencing one of the most incredible sights of my lifetime. *We'll have to go back*, I thought.

After fifteen minutes, the performance sputtered to a trickle. Clicking through my cameras, I nodded out the window, grateful to the universe for indulging my greatest desires. I fell asleep smiling.

Baptism by Fire

I don't need luxury! I thought. It was April 21, 2021, and I was doing my favorite thing in the world: venturing out to the Great Plains for my annual month of storm chasing. That meant booking a cheap motel in Longview, Texas, a stone's throw off Interstate 20.

My yearly pilgrimage to the Great Plains is always a long drive, but this year I had a bit of a head start. Instead of driving all the way from Washington, DC, I had picked up my truck in Birmingham, Alabama, a day prior; it had been there since a tornado outbreak I had chased on March 25.

The night before, I had stayed in Meridian, Mississippi, and after a good night's sleep, I banged out two morning articles and a couple radio hits. Then it was time to hit the road.

This season I was heading out a bit earlier than I had in previous years. Since the pandemic was raging on and most people were working from home, I didn't have to be in the office (or in Washington at all, for that matter). All I needed was a laptop and a (somewhat) stable internet connection. That meant any Applebee's, Chili's, or Dunkin' parking lot could become my de facto workplace for a day. Besides, the *Washington Post* didn't serve BLT croissants or spicy fries, and Chili's did.

Unlike my leisurely drives in past years, I was on a strict schedule. I had to get to Dallas by 8:00 P.M. Thursday at the latest. It was 7:00 P.M.

Wednesday as I approached Longview, a city in east Texas; I'd driven 371 miles since lunchtime and I was in good shape. It seemed an adequate stopping point for the evening, plus the Motel 6 was only $55 a night. I was about to find out why.

~

During severe weather season, things move fast and plans come together quickly. I had only pulled the trigger for this trip on Tuesday, booking a flight to Birmingham for the next evening. That's around the time I texted my friend Gabriel in Boston, whom I hadn't seen in years, but who, in February 2017, had jokingly made me promise to take him storm chasing "someday." I always follow through on my promises, even if years pass before "someday" arrives.

Gabriel had just wrapped up four years at UMass Boston studying psychology and human resources. He graduated a year after me, working in HR at a firm in Boston before transferring to Amazon. In his spare time he raised awareness for social justice campaigns; we were polar opposites, but that made for a good friendship.

Finances were tight, but I happened to have just enough Frontier Airlines miles to score him a good deal on a flight. I was slated to pick him up at Dallas-Fort Worth International Airport around 7:00 P.M. on Thursday.

Allen, meanwhile, had been planning to come storm chasing all along; we had originally talked about a weeklong stretch in May as early as Alaska. But when I texted him that Tuesday night and said, "This Friday could feature a good setup," his response was simple: "I'm in." He had tickets booked within the hour.

I figured both would get along well, and they had conveniently snagged flights that would get them to Dallas at the same time. Twenty-four hours before their arrival, I was sitting at the Waffle House in Longview, Texas, sending them final travel information.

"Make sure you sign the waivers, scan them and get everything back to me ASAP," I texted them. "It's just a formality. You have nothing to worry about."

Sure, it might seem a bit intimidating if your close friend makes you sign a waiver, but when it comes to severe weather, I don't mess around. The second I'm on a chase, I'm all business.

Gabriel had to be back in Dallas by Sunday morning for his return flight to Boston; it would be a quick two-day trip for him. Allen, on the other hand, had plenty of time to spare, so he planned to stay with me for a week. Gabriel's schedule meant a quick turnaround from wherever Friday's chase took us. I knew it was going to be a risky gamble.

For days, I had been eyeing a potential cold-core low setup. A lobe of chilly air and spin aloft was set to slide over the south-central Great Plains. Simultaneously, the low-level jet stream was set to feed into the surface low, and would tug a ribbon of warm, moist air northward, providing unstable conditions that could lead to robust thunderstorm growth.

Weather models were tepid in their simulations; in fact, they were projecting nary a drop of rain. My gut said the models were wrong. *This looks like April 22, 2020 all over again*, I thought. That day had featured a string of pearls, or four small rotating supercell thunderstorms, which crossed Interstate 35 south of Oklahoma City. Three of them produced epic tornadoes. When you do this job long enough, you acquire a gut instinct.

I sent Allen and Gabriel a basic packing list, provided a rough sketch of logistics, and paid for my meal at the Waffle House. Then I strolled to the restaurant parking lot to drive a few hundred feet to my hotel, passing McDonald's, the Knights Inn, the America's Best Value Inn, and a brand I didn't recognize called Express Inn. I'm not a high maintenance fella; as long as it clean and safe, I'm fine with just the basics.

Pulling into the parking lot of the Motel 6, however, I began to wonder if it was all that safe. Empty alcohol nips were scattered along the cement block at the head of my parking space, and occasional whiffs of marijuana floated through the air. That wouldn't ordinarily mean much, but the sideways glances I was receiving from guests as I innocently wheeled my suitcase and camera bag into the downstairs check-in office gave me goosebumps. I had a strange feeling. Something didn't seem right.

Only one person was allowed in the lobby at a time due to the pandemic, which meant a fifteen- or twenty-minute wait. When it was finally my turn, I quickly paid, thanked the clerk, and studied the map on the wall. My room, 237, was in the rear of the hotel on the second floor. I wheeled my luggage back outside, threw it in Ridgie's tailgate, and drove slowly drove around the long side of the hotel.

It was a two-story block-like cement structure. The walls outside were painted manila, with turquoise metal doors and green trim. Since it was an outdoor-entry hotel, walkways with metal railings and banisters wrapped around the second floor. From one of them hung a large red canvas sign that read WIFI HERE!

The back side of the hotel overlooked a Taco Bell and a cemetery. Tall, overgrown trees reached over the fence that separated the properties, as if slyly beckoning me to walk nearer for a closer look. Dusk had fallen, and their leaves all overlapped into an eerie, shadowed web of darkness.

I parked next to the corner of the building, the metal hail cage hanging a few feet out the back of Ridgie's tailgate. In just thirty-six hours, it would be mounted onto the roof. Grabbing my camera bag and suitcase, I scanned around in search of my room, second floor on the left. The sidewalk tiles over which I walked were chipped and misaligned, proving cumbersome as I yanked my three-wheeled suitcase along. *I really need to get a new one*, I thought. My footsteps echoed on the metal staircase as I casually sidestepped dead bugs and beetles.

"Home sweet home," I muttered as I slid the room key into its slot and pushed down on the handle. The familiar scents of unventilated must and cigarette smoke welcomed me into the "nonsmoking" room. It wasn't a lavish room, but it was all I needed. I tossed my backpack onto the faded bedspread, which was dotted with sporadic burn holes. Loose hairs and a few crumbs adorned the floor. A salt packet rested inside the grill of the wall-mounted air conditioning unit.

"Whatever," I sighed, walking over the door to latch the deadbolt. There wasn't one. Having seen the parking lot, however, and with $7,000 in camera and computer gear with me, I wanted some security, so I decided to shove the mini fridge against the door. Big mistake.

"Gah!" I shouted, jumping back in horror. A trio of cockroaches scuttled out from the back of the unit. I hastily stomped on one of them, but the two larger beetles scurried beneath the wall trim. One of their hairlike antennas dangled out, just barely in view.

"All right, Capooch out," I said, promptly grabbing my backpack and reextending the handle of my suitcase. Before I left, I grabbed the free soap and shampoo from the restroom—a little something for my troubles—and tiptoed out the door. I snapped a quick picture of the cockroach-infested floor and tweeted it with the caption "I'm getting too old for this," followed by the laughing emoji. I stomped back downstairs.

"I'm not staying here, and I'm not having cockroaches as roommates," I told the woman at the front desk. "I hate to complain, but I'll be needing a refund."

She didn't even bat an eye. It was clear she'd heard it before.

As I marched back to my truck, I mulled over my options. I was exhausted, but now I had the heebie-jeebies and didn't want to stay anywhere in Longview.

"There's not much between here and Dallas," someone shouted, as if reading my mind. It was an electrical worker in a utility truck.

"Our room was gross, too," he said, explaining he was in the same boat. "We're out of here. Good luck!"

With that, I knew another two and a half hours of driving lay ahead of me. *Oh well*, I thought. *At least I won't have to drive to Dallas tomorrow.*

—

I awoke to the sound of my alarm clock. It was 7:30 A.M. I was at a Hampton Inn in Richardson, a suburb of Dallas, and I had slept like a rock. No creepy-crawlies, either.

I grabbed my laptop from the bedside table and immediately began devouring model data and surface observations. It was a chilly morning in Dallas, barely sixty degrees, but the warm front was lifting north through the region.

By this time tomorrow, it'll be near the Red River, I thought. A few patches of drizzle were accompanying the front, which had socked in the Dallas-Fort Worth metroplex beneath a deck of cloud cover. My forecast thus far was on track. I walked over to the window and opened up the blinds.

"Oooh wow!" I exclaimed, jumping with glee. The sky looked like the turbulent surface of a stormy ocean. The overcast was sculpted into wave-like undulations that stretched as far as the eye could see. I grabbed my camera and ran outside.

I instantly began snapping photos, keenly aware I was looking at a classic display of *asperatus undulatus*, also known as asperitas. They're rare but striking, and sometimes form along warm fronts in the spring months. Displays this dramatic were uncommon, and I could hardly stand to miss a minute. I opened the passenger seat of my parked truck, snagged the waffle and hash browns I had ordered to-go from Waffle House the previous night, and decided to sit on the curb and enjoy a morning picnic.

Asperitas clouds require a stable, stratified atmosphere to form. In other words, air masses have to be comfortably layered based on density.

Imagine filling a fish tank with Italian salad dressing. After a while, all the different oils and waters and vinegars would settle and separate, stacking on top of each other in order of density. The atmosphere on this particular morning was the same.

Consider dropping a stone in one corner of the fish tank. You'd get ripples radiating outward from that point. Now pretend you dropped three or four stones at random. All the wavelets would interact, combining or annihilating one another to produce a chaotic interference pattern. Odds are there were thunderstorms or large sources of vertical motion and rising/sinking air a few hundred miles from Dallas, and I was witnessing that interference pattern firsthand.

After a half hour or so, I figured it was time to get to work. After all, I was pushing 9:00 A.M. Eastern time, and my *Washington Post* articles weren't going to write themselves. I returned to my hotel room to the sound of my phone ringing.

"Hey Matthew, it's Tori from the *Absolutely Mindy* show!" the caller said. I realized it was the final Thursday of the month and I had completely forgotten about my running slot with the popular SiriusXM radio program. Created and hosted by the incredibly bubbly Mindy Thomas, it's a program tailored to educating and connecting with children on their way to school each morning. I was to be a regular.

"What do you think would be a good topic today?" asked Tori, the producer. I thought for a moment.

"Well, I just got to Texas and tomorrow will be my first storm chase of the season," I said.

"Ooh! So you can talk hail and storms and tornadoes? Let's definitely go with that!"

And then I launched in to breaking down the setup I was hoping to chase . . . although I spared the kids the cockroaches.

~

"Follow the sound of the horn," I typed, periodically honking a brief stinger. I was in the Terminal A parking garage adjacent to gate 27. Gabriel had landed an hour earlier, and Allen had just touched down a few minutes before my arrival. I directed them both to my location.

"You'd better hurry up if you want shotgun," I texted Allen. I wanted him in the front seat anyway, and his navigation skills were an added plus.

My phone rang. It was Gabriel.

"Where are you?" he asked.

"Across from A27, parking garage, first floor," I said. I searched the automatic doors next to the loading/unloading zone, spotting Allen from behind an idling bus. I double honked, catching his attention. He smiled, nodded, and began sauntering over.

"Can you share your location with me?" Gabriel requested.

"How do I do that?" I asked. After talking me through it, I successfully sent a pin.

I hopped out of my seat and grabbed Allen's suitcase, tucking it neatly in the back of the truck. He climbed into the front passenger seat.

"I bought you some chips from Piggly Wiggly in Alabama," I said, handing him a bag of potato chips. Piggly Wiggly had come up in conversation when we were discussing funny-sounding brand names.

"Thank you," he laughed.

Gabriel showed up barely a minute later. I took his bag, opened the right rear door for him, and made my way back to the driver's seat. The back left seat was occupied by my mobile hail freezer as well as a tangled mess of wires, adapters, chargers, and cameras. I had spent much of the afternoon preparing the vehicle for Friday's chase.

Allen and Gabriel quickly began conversing, and the three of us headed to the Olive Garden—nothing kicks off a storm chase like bottomless breadsticks. We signed the check and asked to take our breadsticks to go, but the waiter disappeared with them, never to return. It was probably the universe telling me to cut back on my carbs.

~

"So, here's what we've got going on," I said like a teacher, preparing to break down the setup. I was in my pajamas and had my retainer in and sounded like a toddler, but I figured a weather briefing before bed would help Allen and Gabriel understand what may be in store tomorrow. Even if it didn't they were my captive audience.

"We've got a developing low ejecting out of New Mexico, with a lot of cold air at high altitudes," I said, scrolling through a computer model simulation and pivoting my laptop toward them. "That will allow pockets of warm air near the ground to rise and form storms. Changing winds with height means anything that pops will spin."

Allen nodded; Gabriel angled his head and stared intently at the colorful map.

"So are we staying here tomorrow?" he asked.

"Nope," I said. "It's a tricky call. Warm fronts have a bit more twist along them, but I think everything along the warm front in east Texas will merge quickly and become messy. On the other hand, we may not even see a storm if we target the cold front to our northwest, but if one fires, it could be nasty."

At the time, I was torn between the setups. I knew which I would choose, but the pressure of wanting to find something for Gabriel, who only had forty-eight hours in town, was making me reconsider. Did I want high risk/high reward or a guaranteed something that was less impressive?

The cold front had all the dynamics of a big day, but the cap was strong. That cap could suppress surface air and prevent it from rising altogether, leading to a blue sky bust. But if temperatures at the surface warmed above the convective temperature and hit seventy-four degrees, there was a good chance the cap would break.

I decided to put it up to a vote between Allen and Gabriel. I outlined the scenarios and my qualms with each.

"If we weren't here, what would you do?" asked Allen.

"Quanah, Texas, no hesitation," I said. Even though models didn't squeeze any rain out of the cold front setup to our northwest, or even depict any thunderstorms forming, I had high hopes.

"Let's do it," Gabriel replied. "Go big or go home."

I smiled, glad to have subconsciously shifted some of the responsibility for the chase away from myself, or at least I felt like I had. I got back to reviewing incoming data, dissecting every blip, blob, line, and squiggle out loud. Allen, rapt with attention, asked periodic follow-up questions. Gabriel, who had since fallen asleep, began to snore.

—ʍ—

The all-too-routine sound of the default iPhone alarm clock woke me up at 5:00 A.M. It was still pitch dark outside, but I had three articles to write and two radio hits to record. Still in my pajamas, I silently slipped out of my bed and hopscotched over suitcases to a small pull-out desk. I fired up my laptop and got to work.

A rocket had been launched the previous night over Wallop's Island, Virginia, and I knew that Jason would want a piece about it. I began typing that up before shifting to an article about spring flood potential in the mid-Atlantic. By 10:30 A.M., I was putting the finishing touches on a story detailing the forthcoming severe weather threat on the Great Plains, which doubled as an exercise in forecasting for the day. The more intimate I am with the data behind a day's setup, the better.

I pulled out my iPod shuffle, attached an external microphone to it, and recorded my two afternoon NPR Washington, DC, radio forecasts, each beginning with my signature "Well, gang . . ." I figured Washington was a fast-paced, stressful place, and any personality I could inject to make my forecasts conversational would go a long way.

My singsong yammering woke Allen and Gabriel, but it was time for them to get moving anyway. They groggily rubbed at their eyes as I paced around the room restlessly, hastily unplugging camera batteries, chargers, and microphone receiver packs from a half dozen wall outlets. I checked behind the TV and under the bed, yep, I had grabbed those ones, too.

"Bus leaves at eleven," I said, matter-of-factly. It was a little over three hours to Quanah, a town bordering southwest Oklahoma near the Red River, and I knew I had to be there by 3:00 P.M. at the latest. As it was, we were already pressed for time.

I wheeled my suitcase out to the truck, making a second trip back inside for camera equipment. I attached my suction-cup camera mount to the dashboard, clipped my marine weather radio into place on the visor above my seat, and ensured all my cameras were ready to go. Allen and Gabriel appeared as they walked out of the hotel with their luggage in tow.

"What's this?" Gabriel asked, lifting a bundle of items from his seat onto his lap. Allen had a similar assortment of items waiting for him.

"That's everything you'll need today," I said, plugging my phone into its charger and shifting into drive. "The helmets are there for big hail. Once you get to golf-ball size, you're talking big bruises. Hen eggs won't do you any good either, and a baseball or bigger could kill you. If you have to go outside for any reason, wear a helmet. I'll tell you when it's necessary."

"Like, actually?" Gabriel asked. I looked in the rear view mirror. He was raising an eyebrow, mouth ajar.

"Prepackaged safety glasses are there for you to wear too in the event we encounter destructive hail," I continued, adopting the rehearsed tone of a flight attendant running through a scripted safety speech.

"The hail cage should help with that though, right?" Allen asked.

"Yeah, but it's not foolproof. I don't mess around with shattered glass." Both nodded apprehensively.

"Now, lightning is a threat you can't really mitigate. If you're on a storm chase, you're always at a nonzero risk of getting struck. If there's what we call a CG barrage, I'll have us wait inside the truck. And if we come across any downed power lines, we'll do our best to avoid them, but as a precaution I want you to cross your arms and hands and lift your feet of the floor as if you're hugging yourself."

I could tell Allen was a bit anxious. Gabriel, blissfully ignorant, was excitedly composing an Instagram post and taking selfies in the custom hard hat.

"You'll also find a safety information card in the seat back pocket in front of you," I continued. "Actually. Please read it and familiarize yourself with what to do."

I had produced laminated brochures outlining what to do in a variety of hail, wind, flooding, fire, and tornado scenarios. Three different courses of action existed depending on the magnitude of winds. Cliché as it may sound, safety is my modus operandi. If I can't do something safely, I don't do it at all. This applies doubly if I have a guest.

"For low-end winds with debris, recline seat to fully horizontal position, face downwards on seat below window level, and cover head/back of neck with hands," the card read. "Don safety glasses and helmet." Gabriel added the card to his Snapchat story.

"And lastly, you both have an orange thing that looks like a hammer. That's a seat belt cutter that also doubles to smash the windows in the event of an emergency," I concluded. The two remained silent, Gabriel smiling while Allen read through the safety information card line by line.

Gabriel had a work phone call to make while Allen and I bantered about what would happen if a person with the last name Hall donated his money to a university: would they name a hall Hall? Or if someone's surname was Wing and they left a gift for a hospital: Wing Wing?

Harvard had a Haller Hall, which was a relevant point in our deeply intellectual discussion.

Eventually, the need to refuel arose before we had exited the metroplex. I only had ninety miles left in the gas tank, and Allen and Gabriel were hungry for lunch; I dropped them off in front of a Chipotle in Frisco, Texas, while I went next door to gas up.

The ride to my target in Quanah was an uneventful one. We headed west on State Highway 380, passing fields, tractors, and cattle ranches. Small trees, bushes, and shrubbery cluttered the horizon, an impenetrable overcast remaining draped overhead. It was still only sixty-six degrees as we approached 1:00 P.M. I was getting antsy.

Once we got to Decatur, we turned onto Highway 287 north, which connects Fort Worth and Wichita Falls. Allen, in charge of DJ duties, had hooked his phone to the Bluetooth and was playing The Book of Mormon soundtrack. Gabriel and I politely smiled and nodded.

As we crossed through Bowie, I recounted the story of my experience being inside an EF1 tornado as it blew through the town a year before. Like every time I told a story, I sounded like a rarely visited old grandfather in a nursing home weaving long-forgotten tales from his youth. I couldn't tell if either Allen or Gabriel was genuinely interested, but at least they humored me and feigned enthusiasm well.

I eyed the dashboard thermometer: sixty-nine degrees. The clouds were still stubbornly hanging around, but at least the ceiling looked a bit higher up than before. That was encouraging.

Sixty miles east of Quanah, I pulled off the highway in Wichita Falls. I had promised to buy Allen and Gabriel an ice cream cone at Braum's, my favorite staple of the southern Great Plains. While they were ordering, I stepped outside to place a phone call.

"Who's that?" Allen asked, walking outside.

"Nobody," I said.

Except it was somebody: the local Olive Garden two miles up the road. I hadn't forgotten Allen's crushed expression when his breadsticks the previous night disappeared with the waiter, and I'd be darned if he'd be chasing storms without an adequate supply of garlic-buttered breadsticks.

I climbed back into the truck, switched on my all-hazards NOAA weather radio, and drummed anxiously on the steering wheel as I waited for the pair to return. The Storm Prediction Center, a National Weather Service special center focusing on thunderstorms and tornadoes, had issued a mesoscale discussion highlighting the potential for three-inch hail and isolated tornadoes.

Convective initiation, or the rapid development of thunderstorms, was forecast between 2:30 and 4:00 P.M. It was 1:56 P.M., and we had another fifty miles to go. There was only so much time before the cap would break and unleash the atmosphere's pent-up rage.

Once the two had finally piled into the truck, I breezed down the road to Olive Garden, sprinted inside, and exited clutching a bag filled with a dozen breadsticks. I whipped open the driver's-side door, flung the bag of still-warm breadsticks at Allen, and began driving down the road.

"You didn't have to do that," he said, laughing.

"Shut up and eat your breadsticks," I said with a half-smile, my narrowed eyes focused on the road ahead. "You too, Gabriel. It's going to be a long evening."

—∾—

It was sunny. Very sunny. The temperature had spiked to seventy-five degrees, and the sky was deceptively clear. I parked the truck at the first gas station in sight, slid out of my seat, and stretched.

"Welcome to scenic Quanah, Texas," I said to Allen and Gabriel, gesturing with my outstretched hands to a small Main Street behind me.

"Is that where you bought your cactus?" Allen asked, pointing to an adjacent lot crammed with hundreds of metal figurines, lawn sculptures, whirligigs, and rusted antique signs. He was right. A year prior, I had purchased a three-foot-tall metal cactus sculpture from the store; it currently sat in my living room. Now the windmills on the property were whirring, producing high-pitched hum in the fresh Gulf of Mexico breeze.

"Yep!" I said, eyeing the sky suspiciously. With this much of a southerly wind, storms would have no issue drawing in warmth and moisture, or inflow.

"All right, y'all, help me out here," I said.

Allen and Gabriel stood on opposite sides of the truck while I climbed into the pickup bed, my fingers prying at the frayed ropes that had tied the hail gage into place. I hoisted the hail cage over my head, Allen and Gabriel helping guide it forward onto the roof racks so that its chicken wire–paneled protection would be cantilevered over my windshield. They held it steady as I began fastening each of the four attachment joints into place.

"Well, if it isn't Matthew Cappucci!" a voice behind me boomed. The three of us turned. I didn't recognize the middle-aged man, but he seemed friendly; a baseball cap and smile lines make anyone appear amicable.

"Love following you," he said, smiling. "Big fan. I'm James."

I reached down from my perch atop the truck to shake his hand. Before long, we were deep in a discussion of the day's environmental setup. He wasn't a meteorologist, but he had been chasing for a few years and knew me from my work at the *Washington Post*.

A pair of other chasers sauntered up and joined the chat, even asking for a picture. I had always been a nobody, but in this small, insular community, I was apparently a fledgling somebody. Allen shot me a confused glance. I just tiled my head up and smirked as if it happened all the time.

We informally exchanged notes, told our best storm stories, and consoled each other as we shared our biggest busted chases. All the while, I kept an eye on a thunderhead to the west. It had bubbled up and rained itself out quickly, but it signified the cap was almost ready to break. It was only a matter of time.

"Tornado watch is up!" James announced, looking at his phone.

"All right, Larry and Curly, I'm going to run to the restroom," I hollered to Allen and Gabriel. They clearly didn't get the reference. "Y'all wait here and then we're going to head out."

Satellite imagery showed a few heaping cumulus clouds beginning to tower to our west near Childress, Texas, and radar revealed they were starting to precipitate.

When I returned four minutes later, I discovered I had a new hood ornament: Gabriel was sprawled out across the nose of the truck. Allen was standing there awkwardly holding an iPhone and taking pictures.

"I don't mean to interrupt whatever photo shoot this is, but we've got to go," I said. The two of them crammed into the truck, which I promptly fueled up and doused with Rain-X. Then we were on the road.

"Where are we going?" Allen asked. I glanced at the dashboard-mounted radar display on my phone.

"About twenty minutes west," I replied.

It was already evident why—explosive thunderstorm development had ensued, and two or three storm cells were lining up along a dry line, or the leading edge of dry desert air shoving east from the Texas Panhandle. The skies were already darkening, a cauliflower-like tuft blotting out the sun. Overhead, the nascent storm's anvil had replaced the blue backdrop of earlier, pouch-like mammatus clouds dangling from beneath.

The shrill shriek of the emergency alert system screeched over my weather radio: a severe thunderstorm warning was now in effect. As we approached, I could make out a rugged lowering on the left, or southern, side of the storm. A rotating wall cloud was already forming barely twenty minutes into its life cycle.

I shut off the music, silently taking in the storm. It had been a long eleven months, but I was back in my element. Allen and Gabriel seemed excited, nervous, and a bit jumpy.

"Allen, I need a road in the next few miles that will take us at least four farther miles south," I directed.

Minutes later the cabin was shaking as we tumbled over rocks on the dirt country road. The routes were organized on a one-mile grid, etching a lattice across the rural landscape. The storm, still to our west, was visible out the right windows about ten miles away from our position. A yellowish hue occupied the gap between the ground and the rain-free base below the updraft, indicating that large hail was likely falling.

"You know, this thing might go tornadic," I said. I asked Allen to hold the iPhone and focus on the cloud while I narrated as we continued to position south.

As if on cue, my radio squealed again: a tornado warning had been issued.

"Oh my god, are we going to see something!?" Gabriel asked. I chuckled.

"We'll see *something*. I just can't guarantee what," I said.

After another minute or two of driving, I turned right onto a vacant farm road and parked. A barbed-wire fence stretched south of the road, with grassy fields and brambles to the north. I dipped into the breadstick bag, gnawing on one as I concentrated on deciding my next move.

The storm's precipitation-free base, marking the rotating updraft, hovered just to our west, acquiring aquamarine hues as it slowly revolved closer. Flashes of lightning flickered in the sheets of rain falling to the right, marking the separate downdraft of the storm. It was quiet, save for the sporadic peals of distant thunder. The wind was virtually calm, though the grass rippled in waves as loose leaves and detritus rustled on the ground. The atmosphere was biding its time, but it was restless.

So was I. We repositioned east, working to remain ahead of the storm. It was easy, considering the storm was only trucking along at a lackadaisical 20 mph. We rumbled about five miles east until a herd of cows blocked the road.

"Please move," I pleaded, gently honking until they crossed the street.

"Where'd the storm go? Did it die?" Gabriel asked, looking up from his phone.

"No, Gabriel, we turned. Look out the back window," I replied, rolling my eyes and smiling. I turned to look at Allen, who also wore a smirk.

"Oh."

The landscape was beginning to green up as spring built into the southern Great Plains, with many trees resembling bare skeletons of fractal branches just starting to bud. In the distance, the ominous darkened shelf of the approaching supercell hung just above the ground as a haze of rain and hail descended to the north. A few rugged tufts of scud rose into the updraft, a sign of rain-cooled air being ingested into the storm's tightly wound spiral of upward motion. I looked up: a conveyor belt of low-level clouds was streaming north as inflow raced in to fuel the storm. This thing was rotating like a top.

For reasons I still can't explain, I abandoned the storm at 5:39 P.M. It looked great on radar and visually, but was taking its sweet time to produce anything of note. I was getting impatient, desperate to show Allen and Gabriel something and prove myself as a meteorologist. Had I been alone, I would never have budged from my position, but I worried Gabriel was already getting bored.

Ordinarily, I'd chase tail-end Charlie, or the southernmost storm in a line. Usually that storm has the most uninterrupted supply of inflow from the south. But the storm we had been chasing was anchored to the triple point of low pressure, where dry air, moisture-laden air, and cold air all met. The warm front connected to the triple point from the east. It was a classic textbook setup for tornado production, and dozens of storm chasers were parked on the side of the highway waiting for the storm to deliver its main act.

Instead, I drove Allen and Gabriel south into clear air, a brief spritz of sunlit rain falling like a wet farewell kiss from my trusted initial storm. In my gut, I knew I'd regret my decision.

I was aiming for the next storm down the chain: a low precipitation supercell. It was a ten-mile-tall rotating cloud dropping softball-sized hail and little else. There was no rain to obscure the spiral updraft, which would be visible from the south. I figured I could get Allen and Gabriel a photogenic barber pole while waiting for the northernmost storm to get its act together.

It was a bad move. The storm was in the process of weakening, its spin and vertical oomph slowly being cannibalized by the northern storm. That original storm was beginning to bypass us, and I had no radar data. I was unaware that funnel clouds were already dancing beneath the base we had driven away from just twenty minutes earlier.

After roaming around in vain beneath an evaporating cloud for another fifteen minutes, I decided to cut my losses and race back north and try to catch up to the initial storm. The weather radio had been shut off for a while, but when I turned it on, my blood began to boil.

"A confirmed tornado was located seven miles west of Lockett, moving east at twenty miles per hour," the automated Perfect Paul voice announced, as an automated tornado warning update was broadcasted.

"Shoot," I muttered, livid at myself. I depressed the gas pedal and accelerated, sliding along a muddy dirt road as we blasted northward. A rainbow, the kiss of death in storm chasing, arced vibrantly to our east as we tucked in behind the departing storm's rainy backside. Gabriel aimed his camera at the sky. The sun beamed down brilliantly from the west, producing a stunning scene: I hated it.

Allen and Gabriel were quiet, sensing my tension and frustration. No one said a word as we bounced east along County Road 2877. Finally, we emerged onto a paved road, and raced north four miles at close to 70 mph. I was thankful the speed limits were higher than on the East Coast.

As we approached State Highway 70, a typical undivided east-west road with one lane in each direction, cell service suddenly returned. For the first time in forty minutes, I had radar data. My eyes widened at the update: a debris ball was present seven miles to my east, marking where an ongoing tornado was lofting debris so high in the sky that the distant radar was sensing its presence and plotting it as if it was hail. I could do seven miles in six minutes, if we didn't run into any traffic. Maybe all wasn't lost.

Vroom went Ridgie, accelerating aggressively down the road like a plane on takeoff. Allen and Gabriel still weren't saying a word. Finally, Gabriel broke the silence.

"Something you're seeing that we're racing after?" he asked timidly.

"Maybe. Ask me in seven minutes," I grunted.

The roadway was wet, but the sun was bright. For ordinary people, the calm after the storm would mark a moment of tranquility and bliss. I wasn't normal. I wanted to get beneath the dark clouds looming ahead. A vehicle heading westbound passed us, kicking up enough mist to conjure up a rainbow. I scowled.

In the distance about four miles ahead, I noticed a fuzzy vertical line barely visible against the low-contrast backdrop of receding clouds. After a moment, a second emerged to its left, the virtually indistinguishable mass resembling an upside-down triangle. I removed my sunglasses.

"Tornado," I said quietly, jolting the other two to attention. A sleepy Allen snapped out of his afternoon sun-induced trance while Gabriel leaned forward for a better view through the windshield.

"There it is," I said. "We just made it."

Within forty seconds, we had angled close enough to discern the outline of the elephant-trunk funnel, by which point Allen had my camcorder trained on it. I affixed it to the dashboard mount while he fired up an iPhone video and pointed it toward me for a social media post.

"All right, gang, I'm Capital Weather meteorologist Matthew Cappucci," I began. "Destructive tornado on the ground right now over there. We are approaching from the rear . . ."

We watched as the vortex whirled across the road about two and a half miles to our east, the ghostly white funnel seemingly unfazed by my windshield wipers swatting at it combatively. The saturated air meant the twister could condense all the way to the ground, making for a picturesque scene as the stovepipe-shaped maelstrom scoured fields just behind a row of trees and buildings. I was reeling.

The funnel swung to the right, or south, beginning to contract at the surface and widen where it connected to the angry sea of clouds above. The bottom of the funnel became hazy. Within a minute, it faded into oblivion, only a pendant button left aloft where the withered vortex had once hung. Then it was gone.

"Wow, wow, wow," Gabriel said.

"That was so cool," Allen replied in awe.

"Where there is one, there are others," I said. "Tornadoes often come in packs. The storm is going to cycle."

By now, we were pulling into the western side of Lockett, a small town of about 125 people in rural Wilbarger County, Texas. Vernon, about six miles to the northeast, was the county seat. We passed ten or so storm chasers pulled to the side of the road, their cameras still locked on the sky. I was gearing up for round two.

"The southern circulation, which produced the tornado, is dead," I explained to Allen and Gabriel. "I'm expecting a new one to form to the northeast . . . Allen, can you get me three miles due east?"

"On it," he said. After a moment, we came to a fork in the road where Highway 70 curved to the left. He directed me to continue straight.

Corn husks and leaves began raining from the sky, a mystical fall of debris reminiscent of confetti. It was equal parts mesmerizing and unsettling, but in its own way, beautiful and peaceful. The atmosphere was mustering strength for its final act. A new rotating wall cloud was taking shape to our east.

We dodged vehicles on the shoulder of the road, a band of chaser convergence marking the confluence of storm chasers from far and wide. I couldn't fault anyone for wanting a front-row seat to Mother Nature at her most extreme: it was pretty damn awesome. I wouldn't miss it for the world.

Suddenly, I regained cell service, and a new radar frame populated on my phone. I did a double take. We were directly under the hook-shaped curl of rain and hail. The circulation was on top of us, and it could touch down at any moment.

I stayed silent, cautiously remaining calm and in control as I navigated another mile and a half east.

"Let me know if you see any tornadoes near us," I said, focused on the road. Allen and Gabriel chuckled, thinking I was joking. "I'm serious," I replied.

We continued east for a moment until we were interrupted by a frantic shout from Gabriel.

"It's right there!" he yelled.

I turned. Sure enough, a tornado was on the ground a mile and a quarter to our southwest. We were in the perfect position: not too close, not too far. The porridge was just right. I pulled over, gawking at the busy sky.

A sculpted cone funnel hung in the middle distance. The sun was shining, and birds were chirping. The regal tornado stood defiantly in front of clear blue skies. It was wearing a gray scarf. I parked at the end of a gravel driveway on the edge of the road. Aside from the twister, which was sucking up red dirt from open fields, the scene was quaint and quiet.

I grabbed my DSLR camera, yanking the strap that was entangled around my armrest, and eagerly hopped outside.

"You two can get out!" I yelled to Allen and Gabriel, knocking on their windows. "We're safe here."

Reluctantly, they joined me outside, though I urged them to grab their helmets.

"The RFD will be here any minute," I said, referring to the rear flank downdraft, which would swirl a spattering of gusty winds and hail around the tornadic circulation. Supercell thunderstorms are big atmospheric mixers.

Allen recorded me as I delivered a "look-live" report on the tornado, narrating the structure of the storm as the twister posed elegantly for the camera. I gesticulated as if I was standing in front of a weather map in a television studio. Gabriel loaded up on Snapchat material, filming Allen as he recorded me.

"The hail is coming now," I said to Allen and Gabriel, concerned for both their safety and their sanity. "Probably a good time for y'all to get back in Ridgie."

They looked at me for a moment, then silently trudged back toward the truck as I continued to snap high-resolution photos of the funnel.

"And throw on those safety glasses," I said while laughing. I hadn't had time to affix the wing flaps, which protected against wind-driven hail, to the hail cage. I was hoping the RFD wasn't too windy.

Before long, a speck of white flashed past me. Then another. Moments later, the ground resembled a minefield filled with monochromatic popcorn—icy ping pong balls were bouncing everywhere. It was both jovial and destructive. Trees nearby were getting shredded.

The tornado began roping out, stretching into a pencil-thin vortex. The parent cloud was moving east while the ground connection lollygagged. That was the tornado's encore . . . it writhed like a snake until being stretched into oblivion. The sun bathed everything in a whitewash of pure brilliance; it was a meteorological baptism. After a year, I felt rejuvenated.

"Ouch," I said, the hail growing in size. I scurried into the vehicle.

"How big is it going to get?" Gabriel asked.

"I don't think much bigger than baseballs," I said.

"*Baseballs*!?" Gabriel asked incredulously. "Will that, like, cause damage?"

"Oh yes," I replied. "I'm worried about the windshield. I didn't put the wing flaps on."

As if on cue, a loud *thump!* interrupted the rhythmic metallic pings echoing from hail striking the vehicle. The stones were growing in size and falling increasingly erratically.

"There's an egg now," I said.

"How do you know the hail's size just from hearing it?" Gabriel asked. I thought for a moment.

"If I blindfold you and throw a pool ball at you, you're going to know it was bigger than a marble, right?" I asked.

"I guess . . ." he replied.

"I've probably been in more hailstorms than you've been in Starbucks coffee shops," I said. "There's nothing I love more than big hail."

Another large chunk of ice rattled the vehicle. Then another. Gabriel winced as Allen shrunk down in his seat. With every successive crash, my smile grew.

"Because we're in that rear flank downdraft's cold air wraparound, the hail is going to be falling at an angle out of the north," I explained, thinking out loud. That RFD was spiraling counterclockwise around where the tornado had just lifted. "That means we'll get the biggest dents on the left."

"Is the windshield going to be all right?" Gabriel asked hurriedly.

"Meh. Mostly. I want to get to a north-south road so I can angle into the wind," I replied.

Before we could make it there, however, an explosion of ice left a crack in the bottom left of the windshield.

"Nice!" I yelled. "We busted a windshield! That makes y'all real storm chasers now!"

Allen and Gabriel, sensing that I couldn't care less about the now spiderwebbed windshield, seemed to relax a little bit, though Gabriel still clamped his hard hat against his head as if the ice was going to burrow through the roof. He clutched his safety glasses like were in danger of blowing away. I was tempted to remind him that the winds were outside the vehicle, but overall was just glad we were able to get into the thick of it. Our gamble had paid off.

Another loud crackle echoed through the cabin as the lines in the windshield grew. Fortunately, they were relegated only to my side. I cursed myself; I knew I should have installed the damned wing flaps.

"Y'all want souvenirs?" I asked. They appeared visually confused. I glided to the side of the road, donned my hard hat, and slipped outside. The air was refreshing. A thick hailstone pelted me on the back. Ouch!

I ran around for ten seconds or so, scooping up the largest stones I could find like a child collecting Easter eggs. I cradled them in my arms.

"Can you open the freezer, please?" I asked Gabriel, tossing the stones onto the driver's seat so as to not melt them with my body heat. One by one, I smiled at each hailstone like a proud parent and gingerly placed them in the bottom of the freezer. The display read –12 degrees Celsius (10.4 degrees Fahrenheit). An assortment of small stones and raindrops continued to pepper me as I stood nestled between the cab and my open door. Before long, the interior of the door and part of my seat were soaked. Oh well.

Glancing at the radar, I figured the day's show was over. It was now 7:40 P.M., and the storm was gusting out, or becoming outflow dominant.

Since it was exhaling more air than it was taking in, I expected gradual weakening. Only then did I realize how tired and hungry I was. Plus, I was speckled with mud and drenched in rain and sweat, and had somehow ended up with a hailstone down the back of my shirt.

"All right, gang, let's go get some food," I said. "Allen, route us to Altus, Oklahoma, please."

The sun was setting as we cruised north on Highway 283. There were only a few major thoroughfares that crossed the Red River into Oklahoma; the county road networks ended on either side of the riverbank.

To our east, distant thunderheads flickered with lightning. I swore two adjacent clouds were talking to each other, the flashes alternating like a Morse code call and response. In a sense, they were—an electrical discharge in one cloud would be close enough to enhance or change the electrical field in the other. That could trigger or influence a lightning strike. The pale clouds looked like heaps of neatly scooped mashed potatoes.

A new band of thunderstorms had formed to our west along the surface dry line. Unlike their more severe predecessors, which had outrun the front and had long-since moved east, these new storms weren't producing much more than quarter-sized hail. In fact, the developing squall line was thin enough that the sun, just above the horizon, was penetrating through sheets of rain. Only a small sliver of sky was left between the flat horizon and the cloud base above, and the sun was illuminating the cloud from below. I was tired, but I knew I'd regret not getting a picture.

"One sec, guys," I said, turning off the highway and onto the county road network. Train tracks ran a mile to the west, and I wanted a railroad crossing in the shot, partly because of the aesthetic, but mostly because I just like trains.

No one was around, which meant I could stand in the middle of the dirt road. The sky was on fire, a burning orange near the horizon transitioning to a more delicate peach hue that blended with diplomatic purples overhead. The scene was so peaceful and pure that I almost forgot it was a thunderstorm. As I snapped a photo, a sudden bolt of lightning to my north startled me.

"No way," I said, the photo appearing on my camera's LCD screen. I had inadvertently captured the perfect shot: a tendril of electricity was visible snaking its way through the skies right over the railroad crossing, transforming an already sublime sunset photo into something downright surreal. The perfect end to a perfect day.

Of course, that didn't stop me ten minutes later when, driving north, I saw yet another neat feature in the sky. By then, the sun had set and darkness had fallen, though the clouds to the east were glowing a delicate white. Naturally, I just had to get a closer look.

I pulled over and slid off my seat into an adjacent field, which turned out to be filled not with flat dirt, but about five inches of manure. I sank down as if I'd plopped myself into a vat of quicksand. Whoops. (To make matters worse, I was wearing brand new shoes.) Performing my best Cirque du Soleil impression, I used the hail cage to hoist myself into the pickup bed, where I switched out shoes before managing to swing back into the driver's seat from above like Tarzan. (Apparently acrobatics aren't my strong suit, because I would awaken very sore the next day.)

"Y'all are going to love Fat Daddy's," I said eagerly as we approached town.

"Is that a restaurant?" Allen asked.

"Oh yeah. It's the best," I said. "Their chicken sandwich is legendary."

"How do you know these random places?" Gabriel asked.

"When you spend a month on the Plains every year, you start knowing most towns by heart," I said.

It was true: the Great Plains are my happy place, and, in a sense, a time capsule I relish. The wide-open landscape and expansive skies give me room to think, and the slow pace of life offers an opportunity to decompress and focus more on being present in the moment; one can escape it all. There's something comforting about a place that never really changes, where time seems to stand still. Each year when I return, it's as if nothing has changed, providing me an opportunity to reflect on how I have.

We drove down North Main Street into town. By now, it was fully dark, but the puddles on the wide cement roads were reflecting every convenience store neon sign, gas station canopy and headlight. During the day, the city was actually quite quaint.

"You can shut off Susan," I said, referring to the friendly GPS lady. She's accompanied me on tens of thousands of miles and many years' worth of trips, never once yelling at me for making a wrong turn or stopping to stare at the clouds or look up. I aspire to someday have the patience of Susan.

"You know where we're going?" Allen asked, unplugging the phone from its charger.

"Yeah," I confirmed. "Next to the doughnut shop and across from the doughnut shop."

Allen and Gabriel appeared confused, but before they could say anything, I interjected.

"Folks in Altus really like their doughnuts. Happy Doughnuts is my favorite, but Doughnuts and Fried Rice is pretty good, too."

They rolled their eyes and laughed, obviously used to my antics, and my obsession with anything greasy, fried, or bathed in lard.

Like the grilled cheese and bacon sandwich I ordered at Fat Daddy's, and, of course, their signature curly French fries. I smiled at my food, absorbed in the reverie of a perfect day. For the first time, I didn't feel lonesome crisscrossing the Great Plains during my yearly storm chase. In fact, I felt the opposite: I was surrounded by folks who made me happy, including a newfound adventure buddy who would go anywhere and do anything. That suddenly reminded me of my preserved hailstones.

"Oh wait," I said abruptly, grabbing my keys and standing up. "I'll be right back."

I ran to the truck and snatched a few of the larger hailstones, transporting them inside and dumping them haphazardly onto a napkin. Allen just stared at me, while Gabriel displayed his classic puzzled face.

"You conquer a storm, you drink its blood," I said while squinting, giving my best evil laugh. "Mwa-ha-ha-ha-ha."

I plopped a few hailstones into my plastic water cup, taking a big sip and sighing with a contented "ahhhhh." I chucked one into Allen's glass, then turned to Gabriel.

"You too?" I asked, smiling.

"Um, I'm probably good, thanks," he said scornfully.

"More for me," I chuckled.

I scrolled through hotel options on my phone, but it looked like everything in Altus was booked up. Wichita Falls, eighty-five miles to the southeast and about an hour and twenty minute drive away, seemed like the only logical option. We did have to start heading back toward Dallas after all.

We wrapped up our meal and I reluctantly walked back to the truck, less than enthused with the prospect of another lengthy drive. I had promised to type a thousand-word article for Saturday morning's edition of the *Washington Post* offering a chronology of my storm chase, and also had to deliver morning radio hits to WAMU and teach a two-hour morning public speaking class at 8:00 A.M. It was close to 10:00 P.M. I'd be lucky to fall asleep by 2:00 A.M.

The drive was long, but peaceful. Gabriel quickly began to snore in the back seat, but Allen never once shut his eyes. Like on most long car rides, we talked the entire time, speaking softly so as to not awaken Gabriel.

"You're a good copilot," I said, thankful for the company. Heat lightning periodically lit up the tireless storm clouds now a hundred miles to our southeast.

"Thanks for driving so long," he said. I was glad he got to see the tornadoes and witness what it is I do. Storms will always be special to me, but what made it meaningful was sharing it with others, especially someone I cared about.

When the Whole Sky Spins

Things quieted down somewhat after the Lockett, Texas, tornadoes, but there was still plenty to chase. On April 27, a nasty supercell dropped an EF3 tornado on a highway just a few towns away from Lockett; it was rain-wrapped, so I bailed south and didn't core punch like a few others did. Part of me still wishes I had taken the gamble.

April 28 brought us to west-central Texas, where storms spent most of the day struggling. We abandoned them at 7:00 P.M. and drove east on Interstate 20 toward Fort Worth, where Allen coordinated a late dinner with an old friend from college. Our drive was diverted when our phones blared with a wireless emergency alert; to my surprise, a tornado warning had been issued and included the restaurant. A dense core of baseball-sized hail pelted areas on the northwest side of town, but we missed out. (Major hailstorms pummeled San Antonio and Norman, Oklahoma, simultaneously, causing more than a billion dollars in damage.)

By the end of the month, the pattern quieted down. High pressure was in control, and without much to do, we decided to stay in the Dallas-Fort Worth metroplex. I opted to stop burning through PTO and return to writing two or three *Washington Post* articles daily; Allen resumed his thirty minutes a day of "full-time" consulting work. He was still outearning me.

We found plenty to do in Dallas. Against my better judgement, Allen convinced me to ride the four-hundred-foot-tall swings at Six Flags in Arlington. I clung to the chains for dear life, and the wind certainly didn't help. I was mostly concentrated on keeping my eyes shut and not dying, but the scientist in me couldn't help work through some physics equations: it turned out it would take me 4.98 seconds to hit the ground if I fell. *How reassuring*, I thought.

We also decided to check out Dallas's bingo scene, too, heading to the aptly named Betcha Bingo in Irving. The eclectic bingo hall was crowded with bizarre decorations, including human-sized carrot figurines with beady eyes and overzealous smiles that reminded me of Heather; one of the carrots had a crown and a tie, and another was dressed like a farmer. They seemed eerily omniscient, their lifeless eyes peering at us like dolls in a haunted house.

After a game of bowling on April 30, I dropped Allen off at the airport. I dejectedly returned to the hotel room and pored over weather models. Nothing. A bland upper-air pattern meant May would saunter in like a lamb. All I hoped was that it would tap into its inner lion.

It took weeks. Dallas grew pricey, and without much to do, I drove a few hours up Interstate 35 to Oklahoma City, where I found a Hilton that only cost $44 a night. A Twitter fan who worked for Hilton had linked me to his friends and family discount, giving me 50 percent off. If I had nothing to do, a free breakfast buffet with Belgian waffles and whipped cream could at least soften my meteorological melancholy.

On May 8, I drove to south central Kansas to rendezvous with Kelby, who was driving back from her first year of law school at the University of Colorado Boulder. She had expressed interest in storm chasing, so I took her out for a day on the prairie. Storms underwhelmed during the daylight hours, but a hefty hailstorm accompanied us east to a sunset meal at Chili's. We settled into a former Holiday Inn Holidome for the evening, where a late-night supercell delivered golf ball–sized hail and a funnel cloud to our doorstep. That's May on the Great Plains for you.

Allen wasn't set to return until May 16. His 10:00 P.M. flight meant that I had to hang around near Will Rogers World Airport all day. That proved torturous. I watched helplessly on radar as a stunningly sculpted low precipitation supercell thunderstorm produced a high-contrast rope tornado that promenaded for an audience over open fields near Lubbock, many taking to social media to call it the "storm of the year." I couldn't argue: it was. And I was in a hotel laundromat lamenting while washing shirts and eating stale cookies teary-eyed. *This is why I chase alone*, I thought.

—~—

"What the . . . ?" I muttered. It was 6:45 A.M., and I was in my bed groggily reviewing the Storm Prediction Center's outlook. Allen, whom I'd begrudgingly picked up the night before, was still asleep. A sprawling red bullseye had been drawn around the Permian Basin in Texas Hill County, corresponding to a level four out of five risk for severe weather. Jumbo hail was advertised as the biggest threat, the center writing that "some hailstones in excess of three inches diameter are viable."

I got to work frantically writing articles, knowing we had to be on the road by 10:00 A.M. if we wanted a chance of making it to the obvious target on time. A residual outflow boundary from the prior day's storms was draped west to east from just south of Lubbock to near Seymour, Texas. That would bolster low-level easterly winds, enlarging hodographs, a parameter meteorologists use to diagnose the amount of wind shear present in the atmosphere.

Overall, the tornado risk didn't stand out as anything significant. The main hazard looked to be enormous hail the size of teacups or larger. Any chance of a twister or two would be confined to where the invisible boundary became established. Our task would be to pinpoint its whereabouts.

Storms would fire by about 4:00 P.M. ahead of a dry line encroaching from the west. Dew points ahead of it were in the sixties, falling into the twenties behind as desert air sloshed east. The environment would support splitting supercells, storms quickly proliferating and merging into a muddled cluster. We had to latch onto the right storm early in its convective life cycle before it got messy and merged.

I was wrapping up my writing at 9:20 A.M. when the blanketed lump I knew to be Allen showed its first signs of life. I rolled my eyes, powering up my iPod shuffle to record a pair of afternoon radio hits for WAMU back in Washington. If he wasn't awake by now, my hearty, time-honored "well gang!" opening would probably do the trick.

A pair of fluffy waffles and a phone call with Jason at the *Washington Post* later, we were whipping down the H. E. Bailey Turnpike. I was munching on a ten-inch sheet cookie I had bought at Walmart a day earlier, eyeing the sunny skies mischievously.

"What address should I put into the GPS?" Allen asked. He was getting good at his copilot duties.

"Let's do Lamesa," I said. It was about thirty-five miles north-northeast of Midland-Odessa, Texas, a somewhat isolated pair of cities home to about a quarter million residents. I'd been there once before. Empty oil fields surrounded the russet-colored boroughs, a hodgepodge of single-story homes and empty lots playing checkers on the flat, tawny grassland.

"It says we'll get there at 3:14 P.M.," Allen said. I smiled.

"You mean pi o'clock?" I retorted, grinning. I was focused on the road, but I could feel him roll his eyes and smirk.

We passed through Burkburnett around noon, my mind flashing back to the mothership that had angrily gyrated over the Red River barely a year prior. Now the air was purged of its pestilence, blades of grass swaying in the breeze. Wichita Falls followed; eighty-one miles remained in the fuel tank, but I knew never to hit the open road with less than a hundred. (I had learned that lesson in 2018 on the Kansas Turnpike, when I was forced to jog a mile to pay a farmer $20 for a gallon of gas.)

I had allotted time for a quick lunch in Seymour, a small town of 2,700 and the seat of Baylor County. Highway 183 curved around town. I took a turnoff and hopped onto Highway 82, passing the Tractor Shop, Too Wet To Plow liquor store, and Mr. W Fireworks a block north of Main Street. We were in *Texas*, Texas.

"What are you in the mood for?" I asked Allen. Sonic was a no-go, and it was too early in the day for Dairy Queen.

"I could do chicken," he said, pointing to a faded yellow Lego-like building with a gabled red roof and a sign that read CHICKEN EXPRESS. I looked at him skeptically.

"You really want poultry from a place whose logo is a cartoon chicken being violently yeeted . . . ?" I asked, only half-joking.

"I'm down," he said.

We parked and walked inside. The restaurant smelled like frying oil that had been run through a diesel engine. Crumbs and dead flies littered the windowsills, and the plastic overhead menu panels looked like they hadn't been updated since the 1970s. I eyed a piece of stale chicken on a Styrofoam plate that resembled something you'd find abandoned beneath a gas station heat lamp. *No thanks*, I thought.

"I'll be right back," I said to Allen. "I have some data to look at." While he waited in line, I drove a block away to Subway, returning triumphantly ten minutes later with a BLT and a bag of potato chips. I grinned.

"Mine wasn't very good," Allen muttered glumly. I laughed.

"You can munch on your spite cookie on the way," I said. "Let's go."

We made it to Lamesa at around 4:00 P.M. A few cirrus clouds were present overhead and a tornado watch was in effect. The sun was still shining as we parked at McDonald's, but I knew it wouldn't be for long. Several cars were idling in the lot.

"Are you Matt Cappucci?" a voice inquired as I stepped out of Ridgie to survey the sky. I turned. Before me stood a kid a year or two older than I next to a coffee-colored sedan. Based on the SKYWARN stickers on the bumper and pock-marked hood, I hedged a guess he was in town for the same reason.

"Oh, uh . . . yes," I said, taken by surprise. It was the third time someone on the trip had recognized me; apparently folks in storm chase country knew me.

"I'm Andrew," he said. "Love following all your stuff. What are you thinking about today?"

We got to chatting and began comparing notes and reviewing incoming surface observations. Convective initiation was imminent, but we were torn. Should we go north and west to some developing showers along the dry line or farther south and east along the boundary?

"It's tough to turn down those lower sixty dews," I said, referring to the dew point of sixty-two degrees in Big Spring, thirty-five miles to the southeast. Lamesa only had a dew point of fifty-six. It was clear that better moisture resided just south of us. That would translate to lower cloud bases and more energetic storms.

"Good thing you've got your hail cage on," Andrew said. I nodded. CAPE, or instability, was impressive, but was maximized in the hail growth region, or the levels of atmosphere within which large hail forms, usually between about 18,000 feet and 30,000 feet. It was shaping up to be a classic big hail day.

I checked the radar one last time, shook Andrew's hand, glanced at the sky, nodded and walked inside the McDonald's.

"Time to go?" Allen asked, looking up from his McChicken with big, brown eyes.

"Yep," I replied.

Skies were gloomy by the time we got to Big Spring, with pouches of mammatus hanging beneath a new thunderstorm's frayed anvil. We parked on the edge of Intestate 20 for a closer look.

Something wasn't going according to plan. Storm tops were between 40,000 an 50,000 feet, but they appeared poorly organized on radar. Visually, they were nothing to look at, either. The storm was located to my west, placing the updraft on the left; it looked feathery and blurred, a sign of dry air entrainment. This was either the appetizer round or an omen of disappointment.

Temperatures were still in the mid-eighties as Allen and I sat on the tailgate looking west around 5:45 P.M. Occasional loud thunderclaps followed periodic flecks of lightning, with a light smatter of polka-dot raindrops sprinkling the steaming asphalt.

"It's probably a good time for you to head inside the truck just because of lightning," I directed Allen. He obliged. It was my job to protect him at all costs.

A patch of sunshine began to emerge in the middle of the thunderstorm's base, the sliced crepuscular rays painting the dusty sky in a picket fence pattern. The apex of the storm was withering, but the updraft to the south seemed to be consolidating. I racked my mind to figure out what was happening.

Wait, I thought. *It's splitting*. While wind shear was appreciable, it was mainly directional in nature; wind speed was changing with height, but wind direction was relatively constant. That sort of environment favors both clockwise and counterclockwise rotations within storms.

When a supercell divides in a regime of straight hodographs, or only speed shear, the clockwise-rotating left split moves northeastward and produces large hail. The right split, which deviates southeast and spins counterclockwise, has a better shot at tornadogenesis, if it isn't consumed by another splitting supercell. That's because it roams into an environment with warmer, more humid air.

Over the course of twenty minutes, the left split drifted northwest over Highway 87, crossing between Knott and Ackerly well northwest of Big Spring. It resembled a distant mushroom with a clear, sharp base. Allen and I stared at the right split, which, despite pulsing up to 47,000 feet in height, wasn't doing much. Its radar presentation made me yawn, but something about it captured my attention visually. Its base was dark and even, with a well-defined vault forming on the right. That's where large hail would eventually fall.

The northern edge of its updraft seemed to be becoming more well-defined, a retaining wall of grayish-green clouds appearing to hold back the rest of the beanstalk updraft. If it could remain isolated, it would be moving into an environment with better wind shear and more available moisture. The trend was our friend, assuming the storm could remain unperturbed by neighbors.

We drove east on I-20 for a few miles, listening to the dull programmed voice of my walkie-talkie NOAA weather alert radio. My eyes flicked back and forth between the sparsely traveled road and the rear view mirror. A gray filmstrip of precipitation dangled from the cloud against a peachy orange backdrop.

Suddenly, an enormous bolt of light leapt from a towering cloud in front of me. Then another. Within a minute or two, the lightning was virtually constant. A new thunderstorm had developed a few miles to our east, and its isolation piqued my curiosity. I applied my foot on the gas pedal to catch up. Allen turned and looked at me, perceptive that we were shifting targets. He knew me well.

"Want me to reroute us?" he asked.

"You can kill the GPS," I said. "We're just going to wing it."

Doppler radar indicated the eastern cell had exploded in intensity, climbing from 20,000 feet to 40,000 feet in barely fifteen minutes. A wall cloud was forming on its southwest side, which I could see as we closed in on it from the west. I knew that precipitation would soon wrap around the budding circulation. It was 6:06 P.M.

Storm chasing is game of cascading choices, an atmospheric adaptation of *The Price Is Right*'s Plinko. Every decision brings another set of choices—do I target the triple point or the dry line? North or south? Eastern cell or western? Now I was sandwiched in between two storms, and I had a choice to make.

Unlike on previous chases, I wasn't in a rush. Storms were moving at 25 mph, give or take, and even though the eastern cell was receding, I knew I could race after it in a pinch. But I decided to bide my time in between and park on the gavel shoulder of an off-ramp. I wanted to see how things would play out.

That's when the atmosphere decided to complicate things even further. A third cell popped up in between the two, quickly becoming severe and splitting. The sky quickly darkened as the left split approached us from the west. The westernmost supercell was still intact and showed no signs of splitting.

"Ordinarily, I'd go with the lead cell, but I think this left split's going to gobble up that one," I said. Allen nodded, more in support than in agreement. "I apologize in advance if this ends up being a wild goose chase."

I opened the window and stuck my arm out; winds were warm and out of the south, reaffirming my decision. I reattached my seatbelt, clicked it, and shifted into drive rythmically, checking my blind spot before pulling onto an empty Highway 163 heading southbound.

Our road was flanked by sand, small trees, and cacti. The vegetation was just tall enough to obscure my view. With each passing radar scan the trailing, intact supercell to the west was looking angrier. It was beginning to hook. Game on.

"The original supercell is looking good," I said. "We've got to get through this left split though if we want to see it."

A knife could slice through the tension in the air. The trailing supercell was quickly organizing into the "storm of the day," but it had twenty miles to go over open pastures devoid of passable roads. Allen could tell I was stressed. We were so close, but obstacles lay ahead. A mirror-image supercell barricaded our path, challenging us to penetrate its menacing and icy blockade.

Light rain began speckling the windshield, prompting me to twist the knob for the windshield wipers and reach for the headlights.

"Safety glasses," I requested of Allen, who handed me a pair before I had finished speaking. He knew the drill.

The rain became briefly heavy, but we could still see the aged yellow backdrop of the heavily convective sky beneath the baldachin of cloud cover canopied above. Jarring pangs of half dollar–sized hail were a staccato interlude as we approached the clear air ahead of the main supercell.

"In, like, two minutes, you're going to be amazed," I said, confident things would align as I had hoped. I was beaming, every successive radar frame etching my smile a notch more intense.

"Would you mind flipping to the radar?" I asked, nodding toward my iPhone in the cupholder mount. Apple Maps was on screen, and even though the hail had ceased, I didn't want to remove a hand from the wheel to queue up a weather map. Icy musket balls were strewn about the road, conspiring to make the truck lose traction.

EEEEEEEEEEEEEEEEKKKK! gargled the radio through static. I knew what was coming.

"The National Weather Service has issued a tornado warning for . . . northeastern Glasscock County, southwestern Mitchell County and southeastern Howard County in western Texas . . ." the voice proclaimed. I smiled knowingly, feeling both omniscient and clueless. So far, so good.

Skies abruptly lifted as we exited beneath the southern flank of the left-split supercell, affording visibilities of ten miles or more in all directions. A bubble-wrap display of mammatus clouds was draped to the left, their highlights hauntingly beautiful in the colorless evening sun.

"This works," I said, coasting Ridgie to a dirt turnoff that led to a fenced-off driveway. It was May, but fields of bony branches stood atop an emerald-green carpet. The sky over us was veiled by thunderstorm anvils that had all merged into one solid sheet. To our west, the tornado-warned thunderstorm had a scraggy wall cloud. I beckoned for Allen to hop out of the truck, climbing into the tailgate to get a view over the stubby trees.

I closed my eyes and breathed inward in tandem with the storm, buffeted by strong southerly winds racing into the saucer-like cloud. I felt grounded. There's an inherent beauty in the contrasts that make up the minutes before a storm's arrival—the juxtaposition of peaceful light and sinister darkness, the tranquil calmness a premonition of frenzied fury. In just a few short moments, quiet would give way to clamor, the looming storm bearing its tempestuous countenance. It was awakening.

"Look at the striations!" I screamed into the wind. *This escalated quickly*, I thought. I turned to Allen to explain the significance of the tiered updraft. Its layers coiled into the sky like a double helix, the atmosphere's tellurian DNA stripped bare.

It was the first time I could enjoy the start of tornadogenesis without rushing. We were alone with a classic supercell; the atmosphere was exposing its inner workings to us in an intimate and ethereal pageant.

"Is it moving toward us?" Allen asked, gesticulating toward the wall cloud, which now was anchored by a cylindrical cone funnel burrowing toward the ground.

"Yep!" I exclaimed, my happiness unmodulated. I knew the tornado was roaming over fields of nothingness, so I didn't have to temper or regulate my enthusiasm seeing someone's home or livelihood destroyed. Eagerness was coursing through my veins.

"Hold this!" I shouted to Allen, who was standing in the tailgate just to my left. I chucked an 18–55 millimeter Nikon lens at him, which he snatched from midair; I replaced it with my wide-angle lens, hoping to capture the entire supercell in a single frame. "Please," I followed with, realizing I had forgotten to use my manners.

I leapt to the ground and sprinted across the street, noticing a pair of headlights about half a mile down the road. Placing my eye to the viewfinder, I pushed down with my finger. *Click!* It was a bucket-list photograph: one storm, one cloud, one tornado. *Screw yesterday*, I thought. *This is the storm of the year.*

"Is it down?" I hollered to Allen across the street, cupping my hands around my mouth to project. The funnel had dipped down to the tree line.

"I think so," he replied.

The car I had seen moments prior was now just a few hundred feet away. It slowed down and parked behind us. I ran back to Ridgie, exchanging my wide-angle lens for a telephoto lens.

"Yep," I muttered. "It's on the ground."

"Cappucci!" a familiar voice yelled. Against all odds, it was our new friend Andrew.

"I see you trusted the higher dews," I laughed. He smiled.

"This is my first tornado!" he shrieked joyfully.

The clouds above us were racing north faster than I had ever seen before, nourishing the storm as it gained reckless momentum. We were witnessing a textbook diagram that had come to life. Smaller eddies were dancing around the main tornadic funnel three miles to our west, materializing out of thin air and violently thrashing around one another like strands of rope being intertwined into a tortured braid.

"I'm heading south!" Andrew said, waving and jogging back to his car. I could tell Allen was getting nervous, but we had about a dozen minutes before the tornado would cross over our location. We were fine.

The tornado narrowly missed a distant wind farm, the turbines' needle-like arms pointed accusingly at the deleterious denizen. Dust wafted up from the base of the funnel, diffusing into the air like cinnamon in the wind.

I looked skyward. Tentacles of scud were wrapping into the storm. It was psychedelic—we were standing in the middle of an atmospheric merry-go-round. After grinding through Texas rangeland for ten minutes, the tornado began to lift, its tapered shape flattening like a vortex in a cup of coffee returning to equilibrium when it's no longer being stirred.

"You must be a good luck charm," I said to Allen. "I've been wanting that shot for years." He smiled.

"Is it done?" He asked. I shook my head.

"Circulation's going to run us over if we don't get south," I said. "We've got to blast."

We drove south a mile, repositioning as a new funnel started to protrude from the storm's florescent underside. A surging downdraft crashed to the earth around a void-like depression in the belly of the storm, masking the newly minted tornado as the edge of the mesocyclone shifted overhead. The sky was a marble wash of electric green, turquoise, and clementine.

"Would you mind hopping out and holding this?" I asked Allen, handing him my phone. He didn't need any convincing. Seconds later, we were recording an explainer video from the tailgate of Ridgie, illuminating the structure of inflow feeding into the tornado's parent circulation. I didn't require studio lights or weather maps: that's what the sky was for.

A fusillade of lightning sparked fears midway through our explanation, the tornado's silhouette reaching toward the ground like an outstretched finger working to relieve an itch. I asked Allen to head into the truck while I lingered for a last glance before sheets of heavy rain suddenly arrived. I raced inside to rejoin Allen.

"There are baseballs just a quarter mile to our north," I told him as I pinched my phone screen to zoom in on the radar. Dark blue pixels butted against magenta shadings aggressively. "There's a huge tornado in there just behind the rain. It's about to cross the road."

Indeed, it did. At EF2 strength, but invisible outside a small bear's cage surrounding the expanding vortex. I didn't chance it, opting to bail south instead. In storm chasing, it's always better to be safe than sorry. If you make a mistake and play it too safe, you miss out on a few pictures. If you take a gamble and lose, however, it could cost you your life.

The next paved road east of Forsan was about twenty miles to the east, and getting there would entail a circuitous route that would take seventy minutes or more. Sunset was nearing, the storm was departing and radar indicated it had passed us at maximum intensity. I knew the chase was over.

—∾—

An hour and a half later, Allen and I were in San Angelo, Texas, sitting opposite one another in a bustling restaurant booth. I was exhausted, but it was a good type of tiredness: the fatigue of a standout day.

"More breadsticks?" the waitress asked, glancing at Allen and I. Naturally, we were at Olive Garden, and I didn't need to wait for Allen, who had the childlike personality of a dinosaur chicken nugget, to know he'd say yes.

"Yes, please," I said, practically too tired to speak. "And maybe a couple to go. It's been a long day."

Allen smiled. Mission accomplished. *I'm never chasing alone again*, I thought.

When Everything Goes Right

Sometimes, things just go right. Not on accord of any planning, but rather through the universe's favorable auspices. It's true across the board. In the atmosphere and in life, things occasionally have a way of working themselves out.

On June 4, 2019, for example, I drove to Clovis, New Mexico. I wanted to cross the state off my list of places to visit, and I was also predicting some strong late-day storms. A cold front was about to overtake a dry line, and I could visibly see lines of puffy cumulus clouds like pendants lining each boundary by dusk. When the two air masses collided, an explosion of monsoon-like thunderstorms resulted.

The next day was an emotional one. It was time for me to begin the exhaustive drive back to the Northeast. At the time, I hadn't heard back from any prospective employers; it would be a few days before Jason at the *Washington Post* would call. In a fit of whimsy, I decided to delay my return trip for a day, as if somehow avoiding declaring the chase season "over" could defer my future. I settled in Lubbock, Texas, instead.

There was a marginal chance of severe weather, and my hopes weren't high. Weighed down by the burden of my uncertain future and the grueling thirty-two-hour drive ahead of me, I settled in for a nap. Sleep was an escape from reality.

I awoke to a 1930s-style Dust Bowl film. My room was dyed a khaki shade, a bizarre auburn light poking through the windows. I threw on a shirt and rubbed my befuddled eyes, glancing out the Hilton Garden Inn's third-story window. The skies to the west were dark, with a fuzzy dollop of brown clumped over the horizon.

"Haboob!" I thought, frantically donning pants before stumbling outside. I had snagged my car keys and a camera. I didn't expect much from the chase, but something was in the offing. I didn't bother to look at radar, instead driving west past the Krispy Kreme. It took every ounce of self-control not to stop.

Within a half hour, I was face to face with a wall of dust 800 feet high. It would be the first dust storm to strike Lubbock in years; by purely dumb luck, I was front and center to witness it. My weather radio referred to it as an "extremely dangerous situation," warning motorists to "pull aside . . . [and] stay alive."

A stunted line of dark clouds was sliding toward me, dressed in a fabric of dusty mahogany. An instant later, my nostrils were filled, my gritty eyes struggling to bat away the grainy intruders. Visibilities dropped below 200 feet as 60 mph winds heaved through the veldt. My nostrils ached from the granular assault. I never imaged I'd experience a dust storm—or haboob—in Texas Hill Country.

An hour later, I found myself at a coin-operated car wash, aiming the nozzle of a high-pressure hose square at my windshield. I was scrounging for quarters and rubbing my eyes, but it didn't matter, the day had been a success. Sometimes the sky has a way of delivering when you least expect it. I often forget that life has its own timeline. For a person as analytical and exacting as I, blind faith is an unobtainable virtue.

It was a lesson I'd learn again twice in May 2021. Allen left on May 22, but I had more storms to chase. I snagged a weak tornado near Aspen, Colorado, on May 23, and settled in Colby, Kansas, that night. The next morning I woke up anticipating a day off from my pursuit. Apparently I was wrong.

A dingy funereal overcast hung around all morning long, with temperatures around fifty-nine degrees. It didn't feel like tornado weather; I was more inclined to break out a jacket. Despite the rawness to the air, the Storm Prediction Center had drawn a maroon-shaded 5 percent tornado risk around Colby. With a sigh, I decided to hustle my articles to completion, determined to chase locally.

At 1:00 P.M., I was antsy: a tornado watch had just been issued for western Kansas and parts of eastern Colorado. Thunderstorms were about to bust to my south, and I wasn't going to miss out. I hopped in the car, hightailed it into the sunshine just twenty miles away, and basked in the warm, sunny air. The temperature was eighty degrees, and I had crossed into the warm sector of a broader low pressure system.

Little did I know I was in for the toughest ride of my life. Western Kansas had seen a plethora of recent rainfall, and the "shortcut" I had chosen to follow was a muddy Slip 'N Slide–like road with no cell service. For thirty minutes, I white-knuckled it through a gloppy rink of lubricous sludge, petrified to lighten up on the gas pedal. If I slowed to a stop, I wouldn't be able to accelerate again or call for help.

The next four hours yielded little payoff. I found a monster supercell, but for whatever reason it wouldn't produce. Multiple funnel clouds and tornadoes had dropped back near Colby, however, from a storm anchored on the contour of warm and cold. Instead of positioning in the warm air mass, I should have hewed closer to the clashing temperature interface. That's where the boundary had imparted some twist on storms.

Around 6:00 P.M., I called it quits. I was exhausted from a day that entailed difficult driving to no avail, and I ready to return to Colby. Plus, Kelby would be driving west along Interstate 70 from her home in Oklahoma to law school in Boulder, Colorado, and was aiming to meet me for dinner.

En route back to my hotel, I noticed something strange. The lumpy cloud to my north had a strange blue tint; I had never seen something like it before, but it implored me to approach for a closer look. I called Kelby.

"Hey!" I said. "Where are you?"

"About an hour away," she said. We were planning to meet around 7:00 P.M.

"Mind if we aim for 8:00 P.M.?" I asked. "There's a weird cloud that seems suspect. I shouldn't be long," I chuckled.

"Sure," she said. She had an indefatigable patience to which I could only aspire.

As I drifted north on US 83, a sense of urgency swelled within me. The cloud just seemed like it was ready, poised to do something. It was a mere clumping of cauliflower, but there was something it was hiding.

A tornado warning appeared around it on my maps as I neared the intersection of US 24. The chase was suddenly on. The road curved east-northeast, putting me parallel with the storm. I was about five minutes behind it.

A large, royal blue wall hung just to my north, shades of beryl shining prominently. The rain core, presumably on the backside of the storm, was narrow and compact, but it was just a mile north of the roadway. I decided to blast east. A strange funnel appeared through the haze.

As I approached, the funnel came more prominently into view. It wasn't just a funnel, but rather a tornado. Rain was falling, but it quickly vanished as I entered the moat of warm air surrounding the vortex. I was a quarter mile away from fury.

I could hardly believe my eyes—I was among the few people driving east toward a picture-perfect tornado that I hadn't planned for. The entire sky was racing to one counterpoint just a few hundred yards away. *I wish Allen was here*, I thought. A collar of glow-in-the-dark blue wrapped around the voracious vortex.

In a matter of ten seconds, the funnel fully condensed to the ground. It was no longer just a cloudy nipple, but rather an umbilical cord linking the Earth to the sky. I could see sloughs of roofing becoming airborne in the town of Selden just ahead of me. Steel truss buildings were in dire straits.

I pulled to the side of the roadway, eager to document the phenomenon, which was raging barely 1,500 feet in front of me. Precipitation had lifted as I stepped outside, except to my surprise it wasn't cold out. I should be in the cool-air wraparound rear flank downdraft, but instead, the air was warm and dry. I'd later learn that other chasers had, on rare occasions, experienced similarly temperate RFD, which often aids in the buoyancy needed to produce a tornado.

Sixty minutes elapsed before I rendezvoused with Kelby. I was covered in mud and grime and was soaked; my car door was hanging from its hinges, and I had my arm out the window clamping it shut. It had been the best chase day of my life (followed by a rare low precipitation supercell two days later), and I could hardly speak.

While at supper, the restaurant began smoldering orange, as if on fire. Kelby and I rushed outside to behold the setting sun casting an extraterrestrial afterglow on the since-settled scrim of storminess wafting to the east.

It was one of the rare days when the universe had a master plan.

On Cloud Nine

"When do you think they'll come through with the drink cart?" I shouted to Allen over the roar of the wind. It was June 18, and we were in a Beech Craft Turbo King Air 200 flying three and a half miles above Titusville, Florida. It was a stunning morning to take to the skies, but this wasn't a typical Delta Airlines flight: we were about to jump out of the airplane. Allen shot me a nervous scowl, apparently unamused by my joke. I grinned.

Barely three weeks had elapsed since my return from the Great Plains, but I had been feeling wanderlust since arriving back in Washington, DC. I try to have at least one adventure booked at all times so I always have something to look forward to, and nothing was scheduled after storm chasing.

It only took two days of me being home to want to hit the road again. For the first time in my adult life, I decided to take a nonworking vacation, and Universal Studios in Florida was offering cheap tickets. Allen didn't need any convincing.

After reserving passes at Universal and snagging a hotel room in Orlando, I thought back to a conversation Allen and I had while driving through unbounded empty fields in Kansas weeks prior; we had talked about how short life is, and how so few people make bucket lists. Fewer still live by them. It turns out that both of ours included skydiving.

He had laughed when I asked him over dinner in Washington if he actually wanted to try it. "Sure," was the response. Now, at 18,000 feet, he wasn't laughing, but I was. Like a maniac, in fact.

~

The days leading up to Florida proved more exhilarating, and anxiety-provoking, than the actual skydive would be. I had seen an online job opening for a freelance meteorologist at WTTG Channel 5, the local FOX station in Washington, DC. I'd been watching them since I first moved to Alexandria, so, without any real expectations, I decided to apply.

My television aspirations hadn't faded during the preceding two years at a newspaper. If anything, they had grown. I'd delivered live hits from the eye of hurricane on international television and racked up a formidable social media following from my work at the *Washington Post* and WAMU, but getting a paid TV job wasn't looking likely. I began to wonder if my dreams weren't going to pan out after all. Plus, TV is a tough industry to break into, and turnover among meteorologists is very slow.

The United States is broken into 210 different television Designated Market Areas (DMA). Every DMA has its own array of stations, usually one affiliated with each major network, which broadcast over the air via antennas. Most TV stations also partner with cable companies, too.

The markets are ranked on size, which is determined by how many TV homes are present in that DMA. The largest markets—like New York, Los Angeles, and Chicago—may broadcast to millions of homes daily. The smallest only deliver content to a few tens of thousands. Aplena, Michigan, is market 208 for instance; their DMA includes 14,280 homes. New Orleans is market 50, and Detroit is 14.

Pay starts at dirt in the bottom markets. A top earner in market 200 may take home only $15,000–$28,000, which usually comes through working fifty hours a week. Even in markets 70 and 80, most on-air talent work long hours or second jobs to make ends meet. Lots of TV talent are on SNAP benefits. Pay starts to jump quickly in markets 50 and above. More people watching means more ad revenue.

Those DMAs are called large markets, and getting there usually requires several years of experience; some people make it only after a decade or more in the business.

Most students begin their careers in markets 100–210, working their way up in the ranks as the years go on. Climbing the ladder is a process that isn't pretty; parents often have to support their children as they earn near-poverty wages while frequently carrying crippling student loan debt.

The top ten DMAs are considered major market TV; most people never walk into a major market newsroom in their career. The newscasters are polished, veteran journalists and the technology cutting-edge. Washington, DC, is market 7.

I didn't have high hopes when I submitted my application. I had no in-studio demo reel or formal experience, just a cobbled-together montage of social media videos I filmed during my storm chase expeditions. My cover letter basically said "if I can narrate a tornado while being pelted by hail, I'll be fine in front of a weather map." I dared the company to take a chance on a twenty-three-year-old.

To my surprise, I got an email back from Paul, the news director, about a week later, inviting me to a Zoom interview. I ironed my suit jacket, set out my lucky blue tie, and made sure my lighting worked. I dusted a thin coat of makeup on and snaked a coiled earpiece down the back of my shirt before positioning my chair to line up with premeasured marks on the floor. If I was aiming for a TV job, I sure as heck was going to look the part. (I was still wearing sweatpants and flip-flops, though.)

Five o'clock rolled around; it was a Thursday evening, and I had no idea what to expect. Unlike my friends in consulting and tech, I didn't have case interviews to study for or coding exercises to rehearse. I would have to improvise.

"Matt!" a friendly voice boomed, the screen suddenly coming to life. I recognized Paul McGonnagle from a Google search I had performed the night before. He was wearing blue jeans and a striped blue button-down shirt, and reminded me of the classic fatherly next door neighbor you'd see at a New England summertime barbecue. It turns out he was; he had spent years in Sandwich, just two towns over from where I grew up in Massachusetts, and had a son my age.

Alongside him was a man in a polo shirt with curly hair. Paul introduced him as Kyle Carmean, the assistant news director. I smiled, introduced myself and lowered my shoulders. I could tell this would be very, very different from my hedge fund interviews.

The conversation flowed naturally, and without meaning to, I slipped into character, my hands flying as I leaned toward the camera enthusiastically and recounted one of my recent storm chases. Paul asked me about my approach to weather coverage, to which I replied, "You can get numbers from an app. I deliver science and passion." After about twenty minutes, the pair bid me farewell, promising to be in touch.

The next day, Friday, I got an email from the human resources director, asking to meet briefly on Zoom. We chatted for a quick fifteen minutes.

The weekend passed with little fanfare. I met friends for supper, went bowling, and taught public speaking classes via Zoom. I recorded my regular Saturday morning radio hits for WAMU and did a live phone interview about wildfires for a radio station in California. On Sunday night I did laundry, packing a bag for my departure on Tuesday. I also reserved a U-Haul truck: it would be cheaper than a rental car, and we needed a way to get from Orlando to Titusville on Wednesday for skydiving.

Monday brought a hankering for potato chips that could only be satisfied by a trip to the Dollar Tree (they have obscure brands that are tough to find anywhere else). The mission proved successful: they had Zapp's Spicy Cajun Dill chips, bags of UTZ's Crab Chip, and a new type of exotic honey barbecue. I cradled the precious bags in my arms like a newborn baby and walked back to my truck gleefully.

My phone rang. It was a number I didn't recognize, but it started with a 202, Washington's area code, so I answered.

"Y'ello, this is Matthew," I said.

"Hey Matt, it's Paul McGonnagle. You have a minute to talk?"

"Sure," I replied, suddenly interested. I hopped into the driver's seat and sat upright, somehow convinced that my posture mattered for making a good impression over the phone.

"So, I want to offer you the job," he said, "but I just need confirmation you can work any shifts. You'll get plenty of hours. We have a lot of people taking vacation coming up." I was stunned. For the first time in my life, I was speechless.

Nothing prepares you for getting the call that you've been waiting for since childhood. My thoughts instantly became a maelstrom of eagerness and self-doubt. A voice in my head bluntly announced *it's happening.*

"So if I can just get confirmation you can work any shift . . ." Paul said again, followed by a pause. I realized I had been silent for several seconds.

"Oh. Um. Yes," I stammered, suddenly unable to string together a sentence. "Paul, you have no idea what this means. This is the call I've always dreamed of."

"This is going to be great, Matt," he said. I could hear him smiling over the phone. "We can't wait to have you join the FOX 5 family."

The call ended moments later, leaving me to sit processing what had just happened. This was the golden buzzer—somehow I was skipping every rung on the ladder and walking into the big leagues. My thoughts were a tug-of-war between intimidation and self-reassurance. I knew how to talk the talk. Now I had to walk the walk.

I owned two suits and three button-down shirts. All were made in Vietnam when I was at my slimmest (mainly from being a picky eater). I ran through a mental checklist: lose five pounds, give up wine, find a tailor to have more shirts made, and begin memorizing the counties in the Washington, DC, viewing area. I knew Fairfax, Montgomery, and Loudon Counties, but that was it.

I drove back from the Dollar Tree in a trance, somehow missing every pothole and hitting every green light; my truck was floating above the ground. After about fifteen minutes, I remembered to call my parents: my first order of business was to thank them. Everything was falling into place.

Right before I pulled into the parking garage of my building, I called Jason at the *Washington Post*. FOX 5 DC wanted me the following Wednesday, only eight days later. There was no time for a two-week notice, but I also wasn't quitting. Instead, I'd offer them twenty hours a week. It would have to do.

The intense Florida sun was inescapable. It was only 8:30 A.M. as we stood on the edge of a small airstrip outside the Skydive Space Center hangar, but temperatures were already in the eighties. Winds were light, gently stirring the wind sock mounted in the nearby field. The humidity was sultry.

I pressed my index finger against my forearm before lifting it—yep, I was already turning red; it wouldn't take long for me to blister. I began to wonder if I'd be allowed to wear my sunglasses during the jump.

I yawned, exhausted both by the mental gymnastics I'd been putting myself through since Paul's call and from binge-watching TLC the night before. I had introduced Allen to *Thousand Pound Sisters* one night in Mississippi; now he was equally attached to Amy and Tammy. We couldn't pass up a *Thousand Pound Sister* marathon when channel surfing in the hotel, and, naturally, the show made me hungry. A 2:00 A.M. trip to IHOP was the logical remedy.

Now we were huddled awkwardly beneath the shade of a tree as I desperately tried to evade direct sunlight and avoid frying. Other jumpers stared at the ground silently, equally apprehensive about what was ahead.

At about 9:00 A.M., a muscular, tattooed man in a sweat-stained tank top showed up. He had gangly blond hair and an unkempt beard that reminded me of dried ramen noodles. We were promptly handed waivers and told to sign them. Amid the fine print, DEATH AND SERIOUS INJURY stood out. The words were printed in all caps.

We donned harnesses up at 10:00 A.M., the seatbelt-like material wrapping around our legs, torsos, and shoulders. Then we divided into groups of eight. I was eyeing the sky warily, concerned that mid-level clouds emanating from Tropical Storm Claudette in the Gulf of Mexico would preclude our jump.

Skydiving is an exercise in good meteorology and careful planning. Long before any customers arrive or aircraft are fueled, skydive companies pore over plentiful weather data, constructing forecasts by taking into consideration current observations and expected future conditions. Low-level clouds that obscure the ground are a no-go for skydiving. So are cloud ceilings below the height at which the jump is set to take place. Instructors have to be able to see the ground from the plane.

Winds are equally important, not just at the surface but through the entire column of the jump. Strong winds at altitude can blow a jumper wildly off course, whisking them into dangerous territory crowded with trees, power lines, and buildings. The presence of wind shear, meanwhile, can threaten aircraft, causing irregularities in lift during takeoff and landing.

Most skydives, especially in Florida, are scheduled in the morning. By afternoon, the ground heats up enough to allow convection, or pockets of rising air, that can grow into billowy clouds, rain, or thunderstorms. Upward motion within those pockets, which are frequently invisible, can tumble and jerk skydivers or even briefly suspend them. Turbulence can also be disorienting to jumpers.

Complicating matters even more was our location. We were in Volusia County, a stone's throw from the Space Coast. That's prime real estate for the Florida sea breeze, which forms when drier air over the land is heated faster and rises, drawing in slightly cooler air from over the waters to replace it. Most sea breezes kick in by 1:00 or 2:00 P.M. The sea breeze often brings chaotic conditions.

Afternoons can feature a swing from sunny skies to fierce thunderstorms with torrential downpours and earth-shattering lightning displays; it comes as no surprise that Florida is the lightning capital of the United States. Its storms are a daily occurrence that happen like clockwork in the summertime, though Florida's meteorological caprice knows no bounds. In June or July, a late-day thundershower is a good bet by suppertime.

Most skydive jumps originate between 12,000 and 18,000 feet. Below that and you'll have only a brief window of free fall. Anything above 18,000 feet would require supplemental oxygen. Eighteen thousand feet approximates the halfway height of the atmosphere's mass, meaning the air is 50 percent thinner. (Aircraft flying higher would also have to pressurize their cabins, which would mean no opening a door midflight.)

Allen and I opted to upgrade our package to 15,000 feet. We were told to expect about seventy seconds of free fall. We had originally elected to do the 12,000-foot jump, but the tiny, rickety airplane they used for those jumps looked like it had been retired from pulling advertising banners across the sky. Though we'd be jumping from it anyway, I wanted something a little more study.

We were the third group to board the larger twin turboprop aircraft, which meant an hour of sitting in the hangar. I was talking through restaurant selections for lunch and what the rest of our afternoon would entail, mostly to distract Allen, but partly for my own benefit. For obvious reasons, neither of us had eaten yet anyway.

When it was our turn, I folded my hands, said a silent prayer, smirked at Allen, and climbed aboard the aircraft first. Seven other first-time skydivers and their instructors managed to jam-pack the plane, squatting on hard foam benches spanning the length of the roughly twenty-foot fuselage. The final person had hardly sat down before we were taxiing and zipping down the runway—no preflight safety talk or chatter about oxygen masks. We took off eastward into the wind to maximize airflow over the wings. That would boost lift.

I straddled the cramped bench in front of my instructor, who began affixing carabiners to metal loops dangling from my harness. The straps around my shoulders went tight. I glanced out the window as the verdant landscape below faded into a washed-out blue, the curvature of earth becoming apparent as the adjacent coastline shimmered in the late morning sun.

"Is that Cape Canaveral?" I shouted to my instructor while angling my head back toward him, my words swallowed by the growl of the wind. A flimsy Plexiglas door was all that separated the buzzing, vibrating cabin from the vast, empty atmosphere. I estimated we had climbed about three miles in barely seven or eight minutes.

The plane traced a loop back west toward the airfield. Suddenly, a roar filled the cabin as my ears went cold: an instructor had lifted the ribbed Plexiglas door.

"Goggles!" my instructor shouted, tapping me on the shoulder. It was time.

Two jumpers and their instructors took the plunge first. Then it was my turn. I crossed my arms, squatted on the edge of the plane for three or four seconds, and smiled. Then the instructor strapped to my back launched us into thin air.

A peaceful sensation instantly overcame me—a strange calm born out of the complete surrender of control. The acceleration was quick, but it wasn't perceptible; it wasn't like a roller coaster or jet taking off. Within seconds, I was plummeting toward the earth at terminal velocity, roughly 120 mph. An ear to ear grin was plastered across my air-smooshed face.

In the absence of air resistance, we'd continue accelerating at 22 mph every second until pancaking into the ground. Thankfully, Earth has an atmosphere. Terminal velocity occurs when the downward tug of gravity is exactly balanced out by the retarding force of air resistance, which is proportional to speed, preventing further acceleration.

The wind was fierce; even though the air wasn't moving, we were plowing through it. Picture sticking your head out the window on the highway, and then multiply that sensation by five. It was enough to squash my face to look like a pug dog, as if I was braving a Category 3 hurricane head-on.

The air rushed into my nostrils so quickly that it felt like gushing water. Even if the atmosphere at that altitude was two thirds as thick, I was still getting plenty of air. It was crisp and dry; my chilled lips quickly became chapped.

After about twenty seconds of being absorbed in the experience, I lifted my head to scan my surroundings. I spied a strip of land to my east situated just off the coast; a pair of runways were separated by about a mile, surrounded by a row of paved fields flanked by metal buildings. I estimated each to be about the size of a football field, even though they looked as puny as ants. *Wait, that's NASA!* I thought. *Kennedy Space Flight Center!*

My attention turned to the deafening silence that had enveloped me. There were no airplane sounds, birds, or insects to hear, and no one was talking, but I was hearing *something*. I realized it was a whirring hum that was growing higher in pitch and continuously louder.

Ever the nerd, I racked my mind mid–free fall. I quickly concluded it was the product of vortex shedding—little tiny invisible whirls of air were rolling off my face and past my ears. It's the same reason the wind whistles through power lines or your car shakes back and forth when driving behind an eighteen-wheeler on the interstate. The pitch was getting higher as we traveled faster.

Suddenly, I began laughing. *I'm a dropsonde*, I thought, smiling at my meteorological humor. Dropsondes are the opposites of weather balloons, they're small canisters of instruments that are tossed out of airplanes to probe the atmosphere en route to the ground. Most measure temperature, humidity, wind speed, and air pressure. I was mentally logging all those data points during my unimpeded descent.

That's around the same time I noticed the air warming. Air pressure increases near the Earth's surface, since air near the ground is squeezed and compressed by the weight of the atmosphere above it. That describes a premise meteorologists refer to as hydrostatic balance. The compression of air near the ground, coupled with its contact with the surface, is what heats it up.

The temperature had been around fifty-two or fifty-three degrees when we first jumped out of the airplane, but the mercury was quickly rising. I knew that, in dry conditions, the air should warm about twenty-seven degrees for every mile I fell. That corresponded to only twenty-five or thirty seconds of free-fall time. I felt like I was being tossed through multiple seasons a minute.

My meteorological daydreaming was interrupted by a sharp upward jolt as the parachute deployed around 4,500 feet. We were suddenly no longer weightless. Only then did I realize we were up really high. I tightened my grip on the harness. The airport was in view below; a couple other jumpers and their instructors were visible at the same stage in their feathery glide.

Initially, it was smooth, but the ride became a bit choppy—we were passing through the planetary boundary layer. That's the interface between the lowest level of the atmosphere that "feels" the effects of surface friction and a more free-flowing shell around it that makes up the remainder of the atmosphere. The ride smoothed out as we neared the ground, though occasional tugs came when we encountered thermals. Talented parachutists can sometimes surf these narrow columns of rising air, remaining lofted like birds of prey.

Ca-caw! I briefly considered shouting, but Allen wasn't in earshot to appreciate the joke. My instructor would think I was nuts. Then again, it was Florida—anything would fly.

I spend most of my waking hours thinking about the atmosphere, staring at it and studying it; at night I dream about it. Belly flopping through it brought a newfound appreciation for its colossal vacuousness.

Just four minutes after I leapt out of the airplane, I was on the ground bantering with Allen and searching for my sunglasses. But I was still soaring. I had a new leap of faith on my mind, and, like this one, I knew it would take me to new heights.

Home at Last

My first day at FOX 5 DC was on Tuesday, June 22, 2021. I had just returned back from Florida the night before, and only my parents and Allen knew about my new job. I didn't want anyone watching in case I botched my first forecast.

Kyle had scheduled a week of training before my on-air debut, during which time the early morning meteorologist, Mike Thomas, was to teach me the graphic system. I knew a bit about TruVu Max, the software used to generate television weather maps, from a summer internship at the Weather Channel in Atlanta two years prior. The internship had almost scared me away from TV; it didn't help that the station's vice president had referred to me as "Sheldon" in an unflattering reference to Jim Parsons's character on *The Big Bang Theory*. It turns out she didn't appreciate my nerdiness. I hoped FOX 5 DC would be different.

"Matt!" Mike shouted jovially when I arrived at the station at 6:00 A.M. that Tuesday. I had never met him, but had followed him on Twitter for quite some time. I stifled a yawn, wondering how someone could have so much energy before the sun was even up. He walked me inside the cluttered cinder-block building, down a staircase into a musty basement overloaded with boxes and torn-up carpeting. My mind instantly flashed back to Channel 5 in Boston, which I visited at age fourteen; it had the same smell of electronics and humidity. Mike explained that the station was getting ready to move into a brand new state-of-the-art facility down the street in Bethesda, Maryland.

He proved as good a teacher as he was a forecaster. I gleaned that he was only five or six years older than me, which made me feel a little more at ease. Plus, for the first time since weather camp, I had someone to absolutely geek out with. It didn't take long before we were recounting our favorite mesoscale setups.

Mike, whose infectious enthusiasm for virtually everything reminded me of a Labrador retriever, was one of two morning meteorologists. Because FOX 5 DC didn't syndicate any national programming, the station was tasked with producing seven hours of live coverage every morning; Mike handled the forecasting, graphics, and the 4:00–6:00 A.M. on-air time slot. Tucker, twenty years older than Mike and a veteran at the station, worked the 6:00–11:00 A.M. shift.

I had watched Tucker on-air before. He wasn't the classic bubbly, polished weatherman. He spoke casually, conversationally, and intelligently as if without a care in the world, lobbing wisecracks and sideways remarks at the anchors with a mischievous sarcasm that endeared him to viewers. He bled confidence. It was clear he knew his stuff but didn't take himself too seriously. Right off the bat, I could tell that FOX 5 DC was different from the stuffy archetype of a traditional television station.

Meeting him was daunting. He was cool, and I most certainly was not. Plus, he had been there longer than I'd been alive; so had Sue, the station's de facto chief meteorologist, who had worked at the station since 1986. I couldn't wait to meet her; from the stories Mike and Tucker told, she was the station's beloved matriarch.

—

"Dress for air and plan to be there a little before 10," read the text from Paul. Eight days had elapsed and I was feeling good about the graphics system, but my demos in front of the green screen had been shaky. I knew I'd come alive once I was actually on air, but mustering canned sincerity for an audience of zero left me graceless.

Despite my clunkiness, each of the station's six full-time meteorologists spent hours coaching me, offering tips on my delivery, presence, and body language. I was gobsmacked—at virtually any other station, veteran talent would be resentful of a rookie waltzing into a place conventionally out of their league. Somehow this was different. They wanted me to succeed. I got the sense that Paul hadn't been being platitudinous when he welcomed me to the "FOX 5 family" via email. That's exactly what it felt like: a quirky broadcast family.

The station had been planning to introduce me to viewers during *Good Day DC* on July 1; I was told I'd be joining Tucker after the forecast for a brief hello to the audience. Evidently, the station had a surprise for me.

"You're going to be doing the 10:15 forecast," Tucker said matter-of-factly when I showed up.

"*What*!?" I replied, thinking he was joking. He wasn't.

"This hit's on you," he said. "Here. Edit your graphics." He gestured to the computer in the Weather Center.

"I haven't even looked at the weather today!" I said indignantly. It was 10:07 A.M., and I'd just been informed I'd be marching in front of the green screen on a potential severe weather day. Tucker seemed amused.

"You know the forecast," he said, smiling reassuringly and slyly. Deep down, I knew he was right.

Karen, the floor director, handed me an IFB box to connect to my earpiece so I could hear program audio. I snaked a lavaliere microphone down my shirt and clipped the transmitter to my belt.

Here goes nothing, I thought, frowning.

—∞—

An hour later, I was on cloud nine once again. It went great! I stumbled when I was talking about myself, but as soon as I segued into the forecast, I was in my element. It helped that the afternoon was looking spicy, which gave me plenty to talk about. I even went so far as to mention rotating thunderstorms.

I drove home at 11:00 after *Good Day DC*, arriving to my apartment in Alexandria, Virginia, around 11:30. I glanced at computer models suspiciously before settling in for a nap. I could tell the setup wasn't run-of-the-mill.

FOX 5 DC had asked me to be back that afternoon to make a brief appearance in the 4:00 P.M. and 5:00 P.M. shows, as well as join a trio of anchors for *Like It or Not*, a pop culture commentary show that aired every day from 7:00 to 7:30 P.M. That was the one thing I was apprehensive about. Atmospheric fluid dynamics were easy; keeping up with the Kardashians had never been my strong suit.

High-resolution weather model simulations had been painting a splotch of storminess along the Interstate 66 corridor and directly atop Washington, DC, and Route 50 during the afternoon. I guessed that there was some sort of stalled boundary, albeit subtle, that the models were keying into. That's where cells would fire first, and linger throughout the evening. I decided to head back to FOX 5 DC early around 3:00 P.M.

I parked in the parking lot and grabbed my lucky blue tie, scanning my ID badge at the door. *I can't believe it worked!* I thought. I bounced down the stairs, trod to the rear corner of the basement, and slipped off my backpack. A woman in a crisp red dress with short blonde hair and bright cheerful eyes, who appeared to be about sixty years old, was sitting at one of the two workstations.

"Matt!" she said, standing up and extending her arms to give me a hug. It was Sue. I smiled, hugged her, and introduced myself.

"I'm so excited to finally meet you!" she said.

Caitlin, the other evening meteorologist who alternated shows with Sue, was working upstairs in the in-studio weather center prepping for the 4:00 P.M. newscast. Sue's first show wasn't until 5:00, which gave us time to chat. She asked me about all my adventures; we scrolled through storm-chase photo albums on my computer, relived wild trips I'd had overseas, and discussed our backgrounds in weather. It was as if we had known each other for years; I got the sense that was why everyone loved Sue. She was a ray of human sunshine.

We were casually keeping half an eye on the desktop computer monitors, babysitting the radar as thunderstorms formed. I had GR2 Analyst, a high-tech radar visualization software, on my laptop. Sue slid her desk chair over to peek at the display. She asked me what my thoughts were.

Did Sue just ask me *my thoughts?* I thought. It was a feeling I hadn't had since that first weather conference I'd presented at when I was fifteen. Sue wasn't just being nice (she was too genuine for that); she actually was interested in my opinion on the storms. She wasn't treating me as a rookie or newbie, but rather a respected colleague.

"That boundary has me a bit intrigued," I said to her, sharing my concerns that storms would "train" along it and produce flash flooding. Wind fields weren't overly impressive, but a stagnant boundary could locally enhance low-level spin enough to be problematic. We knew Caitlin was on it, so we went back to discussing Sue's recently born granddaughter. I could tell Sue would be an amazing grandmother.

About ten minutes later a red box popped up around a cell just east of Washington, DC. I double-checked on RadarScope on my laptop to make sure I wasn't seeing things: it was a tornado warning.

"Tornado warning!" I said to Sue slightly louder than I had anticipated, sitting upright in my chair eagerly.

"We'd better get upstairs," she said. *We?* I thought. *Oh yeah. I guess I work here, too.*

I sprinted up the stairs two at a time, Sue appearing moments later. Caitlin was already on the air for breaking wall-to-wall coverage. I propped my laptop down at the anchor desk, ready to watch Sue and Caitlin in action.

"Here's your mic," a voice from behind me whispered. It was Maurice, the floor director, who was handing me a wireless microphone and an IFB box. I was confused.

"Sue asked for you to be on air, too," he said. I was shocked and surprised; the chief meteorologist (who had humbly told me, "I'm not a chief. I'm just Sue who works evenings" with a smile) wanted me on-air alongside her and Caitlin to walk viewers through a tornado warning. My eyes practically teared up; I was touched. Sue walked into the studio, nodded at Caitlin, who was mid-sentence, and smiled at me.

"We're joined by our newest meteorologist, Matt Cappucci," she said. The three of us were off-camera, narrating the radar, which Caitlin was driving as it looped on screen. "Matt, what are you seeing?"

My TV career had been going for a whopping five and a half hours and I was already covering a tornado warning. I had been joking for weeks that my first day would feature a tornado. The atmosphere was delivering.

"Thanks, Sue," I said before diving into what had caught my attention; I mentioned that the storm, which was about to cross the Route 50 Bay Bridge over the Chesapeake, was going outflow dominant. That meant strong gusty winds to 70 mph, but a lesser tornado threat. I could tell Sue valued my input. We continued on the air for a half hour until the warning for Anne Arundel County expired. It was exhilarating.

The following hours were comparatively tranquil. After my routine hit at 4:45 P.M. and foray into trending news on *Like It or Not*, I loosened my tie and unclipped my microphone. *Day one. Done.*

The station's head anchor, Jim Lokay, invited me out to supper to mark the end of my first day on air. We had become acquainted via Twitter during my time in Washington, but our paths had actually crossed when

I was a high school senior visiting Channel 5 in Boston. He was working there at the time. I was too starstruck then to say hello or introduce myself. Nine years later, apparently, I'd have a do-over. I was still nervous.

Jim had me follow him down the street in my truck to a pub called Clyde's, about a half mile from the station. Toward the end of our meal, he excused himself to answer an email. I seized the opportunity to check the radar on my phone. With a start, I realized that we were under a severe thunderstorm warning. A gnarly storm was minutes away from our location.

"We should probably get out of here soon, or else we're going to get soaked," I told Jim. He paid for the meal, I thanked him, and we headed our separate ways, Jim back to the station to anchor the 10:00 P.M. news, and me to my apartment building to retire after a successful day.

The sky had an odd look to it as I drove down Wisconsin Avenue in Bethesda; it was 8:30 P.M. and dusk had settled in, but incessant flashes, as if from a strobe light, were casting flickering shadows behind trees, streetlights, and traffic signs. The highly electrified thunderstorm was on our doorstep. As I eased Ridgie to a stop at a red light, I tapped the RadarScope icon on my phone.

I did a double-take as soon as I saw the radar plot: a sinister blob of purple appeared within the storm, which had acquired an arcing kinked shape. A slender red box was drawn around it for downtown Washington. *A red box!?* I thought. Tornado warning. Another storm was feeding off the boundary.

The FOX 5 DC station was just a block away; my shift had long since ended, but it had been years since the nation's capital had a tornadic thunderstorm aimed directly at it. I raced for the station like a bat out of hell, simultaneously parking and tightening my tie before sprinting inside. The first sheets of rain rode in with a curtain of fog just as I reached the door.

Bursting into the studio, I could see that Sue and Caitlin were already manning their stations and had broken into scheduled programming.

They were knee-deep in wall-to-wall coverage. A quick glance at a new radar scan revealed a classic QLCS kink in the line circulation. I knew it was about to produce.

Much like earlier, the floor director handed me a microphone; I still had my earpiece in from earlier. I looked over at Sue, who caught my eye, smiled, and nodded. Caitlin was reciting a list of towns in the path of the storm, which had just produced an EF2 tornado.

Rolling up my sleeves, I sauntered over toward the green screen, awaiting my cue. The weather map looked like a perilous oil slick.

"Matthew is back, and we thank him for extending his day to be with us," Sue said to the cameras. "Matthew, what do you see?"

After a decade of feeling homesick for a place I had never been, I felt like I was exactly where I was meant to be.

Home at last, I thought.

In My Own Backyard

It took about a month, but by early August 2021, I had gotten my sea legs at FOX 5 DC. I was comfortable on-air, had developed a rapport with the anchors, and had familiarized myself with the rundown of each show. I still had plenty of room to improve and become crisp, but my imposter syndrome was starting to wear off.

With every passing day, I came to realize that Paul and Kyle were the antithesis of traditional TV bosses. Unlike the staunch, micro-managing news director mold I'd mentally prepared for based on the tales of colleagues elsewhere, the pair were laid back and hands-off in their approach. In fact, the whole station seemed that way. Counter to my expectations, no one ever tried to tamp down my geekiness. If anything, they embraced and marketed it.

FOX 5 DC's Creative Services Department orchestrated a promo, or fifteen-second commercial, promoting the weather department. They nicknamed it "the translator," and, to my surprise, I was asked to appear in it. I had never heard of freelancers being included in promos, and it turned out no one else had, either.

"This is a great sign for your future here at FOX," Sue texted.

Creative Services opted for a skit that began with Jeannette, the station's lead morning anchor, asking me for the forecast, only for me to bombard her with jargon and buzzwords like *omega* and *vorticity*.

"Tucker?" she asks, visibly confused.

"It's going to rain, Jeannette," he decodes bluntly without ever looking away from his computer monitor.

"That's what I said," I whisper glumly before Mike appears on-screen to offer reassurance for my rookie mistake. The promo turned out to be hilarious and a hit with viewers. *I guess the higher-ups do like me*, I thought.

The summer also brought another couple tornado warnings, both of which I happened to be at the station for. One storm on July 29 dropped tennis ball–sized hail near Fredericksburg, Virginia. It was my day off, but Kyle had asked me to come in and dress for air. Sue, Caitlin, and I tag teamed nonstop coverage, during which I dissected the storm's hail scatter spike and correlation coefficient signature while live. After covering another tornado warning in the Maryland Panhandle on August 18, I could tell Paul and Kyle knew tornadoes were my thing.

I was still working for the *Washington Post* as well, and had recently begun producing video content for MyRadar, an extremely popular mobile weather app with fourteen million monthly users. Andy Green, the CEO, had reached out a year prior to inquire about collaborating, but my full-time role at the *Washington Post* meant external involvements were subject to red tape and bureaucracy. Fortunately, my new contract gave me unfettered freedom, and MyRadar was another opportunity. I was now working on TV, radio, in print, and on an app.

Andy was a genius who never let his inner child fade; he was as into tech as I was weather, and his Orlando-based office was a shrine to his passions: a replica of R2D2 stood in the corner, with an original 1987 iMac on display in the hall. It was a video gamer's paradise. A MyRadar party bus was parked in the lot outside the building, and the coffee table in the lobby was filled with sand that had been sculpted by a marble driven by concealed magnets. *I want to be like him someday*, I thought after our first meeting.

Andy had built an empire decades earlier as one of the first internet service providers in Rhode Island before shifting to new endeavors. Among them was launching MyRadar; he created the app just for fun, but it took off like wildfire. It didn't take long for him to amass a team of thirty people. The job title he uses to describe himself on Slack is "pointy-haired boss." He can usually be found sporting a smile and loose-fitting tank top.

Originally, Andy had wanted me in Florida, but he decided to let me work remotely from Washington after I shared how much the nation's capital felt like home. After all, that's where my adventure buddy lived. Plus, I wasn't about to depart FOX 5 DC. For now, we were experimenting with me doing some storm chasing and weather explainer videos, and sending the footage to editors in the Orlando office.

I chased Hurricane Henri into Connecticut for MyRadar in late August, but the storm weakened upon final approach northeast of Long Island. Aside from some 60 mph wind gusts, I didn't have much to show for my time. A chance to redeem myself arose when Hurricane Ida formed south of Grand Cayman Island on August 26; every meteorologist knew it would spell big trouble.

Ida was drifting toward the Gulf Coast, where high pressure at the upper levels would aid in the storm's high-altitude outflow. There was also a loop current of near ninety-degree waters in the Gulf. Even the most conservative weather models pegged Ida's odds of rapid intensification at five to ten times normal. Louisiana was staring down the barrel of a monster.

Despite MyRadar's willingness to foot the travel bill and the forecast for Ida to approach Category 5 strength, I couldn't chase it—I had to work the weekend evening shows at FOX 5 DC on Friday, August 27, as well as August 28 and 29.

During the first of my three-day stint, I took the unusual step of mentioning the chance for tornadoes and flooding in the mid-Atlantic six days in advance. Ida's remnant swirl would be working up the Appalachians, and unlike previous storms, my gut told me this one was different.

"One thing we have to pay attention to here in the mid-Atlantic . . . is the potential for heavy rainfall and a few tornadoes mid to late next week," I tweeted. I made sure the seven-day forecast on FOX 5 DC had an angry thunderstorm icon for Wednesday, September 1. *I'm really going out on a limb here*, I thought.

That Sunday night, I watched with awestruck horror as Hurricane Ida's black, empty eye swirled on satellite toward the Mississippi River Delta, explaining for viewers in Washington that the storm had strengthened at triple the rate needed for rapid intensification. A gust of 172 mph had just been reported in Port Fourchon, Louisiana, and most of New Orleans was without power. A large part of me felt like I was missing the action, but I was secretly relieved to be safely in a TV news studio a thousand miles from the carnivorous storm surge.

I drove home after the 11:00 P.M. news on August 29 purely exhausted. It was my second fourteen-hour day in a row. Between working on live update files and pitching in with storm coverage at the *Washington Post*; fabricating diagrams, tweets, and explainer videos for MyRadar; and taking calls from international TV and radio outlets, I was busy around the clock. Yet the longest days were ahead.

By August 30, the scope of the damage in the Deep South was becoming evident to the media, meaning my phone was ringing off the hook. I was still working at capacity for the *Washington Post* and MyRadar, but I had become the de facto US weather correspondent for outlets like Germany's DW News, Sky News Arabia, CTV in Canada, and BBC World News. I was happy—after all, I was making $80–$200 for each four-minute hit, but keeping track of the various time zones was depleting. I considered waking up at 1:00 A.M., 2:00 A.M., 3:00 A.M., 5:00 A.M., and 7:00 A.M. and sleeping in an ironed shirt as a typical day at work.

Another round of promos, this time photographs, were slated for Wednesday, September 1 at FOX 5 DC. I emailed Kyle the night before, letting him know about the severe weather threat and hinting that I'd like to work after my shoot. He told me they didn't need me, which was true; all the meteorologists had their promos that day, so it wouldn't make sense to keep a freelancer around, too. I wasn't happy about the decision, but I understood it.

Wednesday morning came early: at 2:00 A.M. to be exact. *Is it the Fourth of July?* I groggily wondered. Nonstop lightning was leaping outside my window with a pair of rotating supercell thunderstorms riding a warm front into the Washington, DC, metro. *That looks borderline tornadic*, I thought incredulously while looking at radar in the dead of night, but both storms remained under wraps. If supercells were firing at 2:00 A.M., I shuddered to think what 2:00 P.M. would bring.

I was up at 6:30 A.M. writing for the *Washington Post*, authoring two quick articles, including one about Washington's afternoon tornado threat. I felt like a Great Plains forecaster on "one of those days." I had awoken tired and knew an eighteen- or twenty-hour day stood ahead of me. *You know you love this*, my mind seemed to say.

The other piece I wrote was about the propensity for crippling rainfall in the Northeast. Hurricane Ida's moisture would pool along a cold front, which would ring the humidity out of the air as if from a waterlogged washcloth. The cold front, the same that would spark severe weather, was parked parallel to Interstate 95. That's never something you want to see.

After both articles and a few radio interviews, I ironed my suit while doing a Twitter Space with MyRadar reviewing the high-impact forecast. Midway through, I received a message from the office of the Montgomery

County School District's superintendent. They were deciding whether or not to release students early ahead of the afternoon tornado threat. I offered my recommendation and applauded their forethought. I was still reviewing weather maps and conducting my mesoanalysis as I hopped in the shower.

I arrived at FOX 5 DC around 12:30 P.M., an hour and a half before my promo shoot. I had a feeling I'd be needed early. The sun was shining through patches of low-level cloud cover. It was already warm and humid—the presence of sunshine was about to help some kernels of major instability to pop.

Tucker and Mike were on-air with the day's first tornado warning by the time I got inside. It was eighty miles south of Washington, DC, on the southern flank of our viewing area, but it was a sign of things to come. I knotted my tie, painted on a layer of makeup, and casually strolled by Kyle's office.

"Any chance y'all need an extra hand?" I asked, smiling.

"We have too many mets here as it is," he said, reiterating the point he'd made via email. I frowned and nodded. I knew it wasn't personal, but I felt like an athlete who had just been benched. *I'm the tornado guy!* I grumbled inaudibly. My shoulders slumped as I walked into the green room to see if I could have my picture taken early.

"You know if Caitlin's coming?" the photographer asked. "It's her turn." He was sitting in a chair thumbing through photos he had taken of the other on-air talent.

"She's in the studio with a tornado warning," I said. "Mind if I take her slot for now?" I sauntered past metal stands suspending umbrellas, reflective padding, and external flashes before posing in front of a white background. Twenty minutes later I was sprinting out the door clutching a bent metal coat hanger and a balled-up white button-down shirt. Half-removed makeup was smeared across my face. The untied shoelaces on my tattered sneakers were flying unrestrained.

"You guys want some chasing today?" I texted to MyRadar's lead meteorologist, hopscotching across the street between cars. I didn't have time to wait. "Sure!" came the response. MyRadar was willing to match my hourly rate from FOX 5 DC. I was determined to do tornado coverage for someone; I had a strange feeling that I'd be needed before long.

I opened the MyRadar app on my phone and flicked through a surface map as I skipped down a flight of stairs into the parking garage. *I've got to get to Anne Arundel County*, I thought. Temperatures were a degree or two warmer east of Washington, and a batch of rotating storms was in the process of blooming to my south. The warm waters of the Chesapeake Bay would also boost tornado chances. The air would be more unstable and cloud bases would be lower.

It was now 1:30 P.M. and skies had clouded over. Sheets of rain were pouring down in two-minute intervals as I poked through gusty squalls en route to the Beltway. My phone rang as I weaved between lanes, convinced I'd choose the fastest one. Washington drivers are not equipped to handle torrential downpours.

"Hey, Matt, it's Gwen." Gwen was the weekend morning meteorologist who had coached me with the same dedication a parent would their own child. She was strict, patient, exacting, and, above all, kind. I hadn't seen her since my first day of work.

"Hey, Gwen! What's up?" I asked cheerfully. I was fully in tornado mode, but was eager to catch up with her.

"Are you at the station?" she asked. "I'm supposed to be heading to where there might be damage later. Any thoughts?" I advised her to head toward Bowie and be prepared to move. As my urge to blast east increased, the traffic in front of me seemed to slow down.

At last I made it onto Route 50 and was racing toward Annapolis. The sky was a grayish yellow. Radar now showed an area of intense, albeit diffuse, rotation about fifteen miles to my south-southeast. I wanted to get east of it in case it further intensified.

Tropical tornadoes are notoriously difficult to chase. They're usually wrapped in rain and only touch down briefly, cloaked from view until it's too late to hide. Their circulations are shallow and tough to spot on radar. Unlike with traditional supercells, I couldn't approach from the south or from the rear: I had to get ahead of it.

The area of rotation began to tighten ten miles to my south. It would be close; if I didn't get off the highway soon, I'd end up on the roadway with a mesocyclone bearing down on me. The cars in front of me had slowed to a crawl. We were now in heavy rain streaming northeast of the circulation.

I was three miles away from my exit at 2:10 P.M. when the rain suddenly stopped. Above me was a jet-black cloud that seemed to wrap to my south in a big arcing curve. I passed a clearing on the right and turned my head—a column of darkness connected the cloud to the ground. I frantically dialed Gwen. There was no time for me to question the absurdity of what I'd seen.

"Cone tornado near Annapolis," I said, out of breath. "It's on the ground." I hung up abruptly and turned off the exit. Three minutes after, I had made my way to Navy-Marine Corps Memorial Stadium.

Should I run up the bleachers? I thought. The tornado was a couple miles to my west, but I couldn't see its snout below the tree line; instead, the rugged, cylindrical wall cloud appeared like an amorphous cylinder gyrating along over forests. I parked and leapt atop my truck's roof for a better view, texting Mike and Tucker "tornado on the ground" before filming a Twitter video for MyRadar.

"I can tell you, that very well may be it!" I shouted at my iPhone, adopting an uncharacteristically grave tone "Edgewater, circulation's over you *right now*. Everybody in Annapolis needs to be in shelter *right now*." "Likely tornado in Annapolis!" I typed.

An eternal minute passed as I contemplated my next move; I was witnessing a historic event and had to get the best vantage point I could.

Without a strong knowledge of the area, I decided to race northwest again on Rowe Boulevard toward the highway. I cursed at the traffic lights obstinately impeding my progress. A mother and her young son obliviously entered an orthodontics office to my right. *Do they not know what's going on!?* I thought.

I passed a church on the left and a globe-like water tower; coupled with the jade-colored leaves whimpering in the wind, the scene was reminiscent of my jaunts through Mississippi and Alabama. An ornate brick sidewalk flanked the roadway. I was in the passing lane.

My eyes fixated on the cement road in front of me as I approached an opulent artisan-constructed bridge over Weems Creek. The sailboats to my right were still, refusing to acknowledge the weak breeze. To my left, a gray pillar of roiling smoke was churning up vegetation.

Wait a second, I thought. I turned left, not believing what I was seeing—it was a tornado, and I was barely 800 feet away from it. It was paralleling me on the roadway. I grabbed my iPhone and began filming out the window, screaming, "Tornado in Annapolis, Maryland!" It was exactly what I had predicted, and the last thing I had expected.

There was only time for split-second decisions. Should I keep going? Stop? Pull over? The tornado was about to cross the road in front of me. A thicket of trees was present to my left, and there was no breakdown lane or shoulder to pull off of. I accelerated before braking in a vacant turn lane, coming to a stop and swiftly exiting the vehicle.

"STOP!" I screamed at cars unwittingly driving into the circulation. A swarm of leaves was floating out of the sky like a morbid snowfall, complete with pieces of shrapnel and tufts of insulation. It was a supernatural beauty, the spectacle a treaty of harmony transfiguring the storm cloud's annihilative appendage. Shreds of vegetation were gliding through the air like feathers, twinkling as they swiped through fissures of sunlight that were pouring through cracks in the threatening sky. The atmosphere was trying to seduce me closer.

A sharp mist swallowed the road two hundred feet to my northwest, where clumps of foggy condensate were rapidly wafting up into the sky. A garden hose of rain sprayed me as the rear flank downdraft wrapped southeast with a blast of cool air. Brake light shone beneath the green exit sign marking Route 50; it was 2:23 P.M. The tornado appeared to be lifting; I checked for incoming traffic, slipped into my vehicle, and prepared to drive once again. "TORNADO IN ANNAPOLIS!" I tweeted, tagging MyRadar.

Concentrating on my rapid breathing once more, I chuckled. *Only me*, I thought. An hour earlier I had been wearing a suit and tie in a cozy TV studio; now I was covered in torn-up pieces of leaves, drenched in water and sweat, and was sliding around on a soaked car seat. *THIS is my natural habitat*, I thought. I didn't have plastered-on smile or the stock persona of a typical on-camera newscaster. I was a weather nerd who was right where he was supposed to be. I needed a towel and a good shower.

I navigated to the highway once more and drove east over the Bay Bridge, half-glancing at radar. My cell phone rang. It was Kyle.

"Matt, we need that video," he said. He must have already seen it on Twitter.

"Let me get permission from MyRadar," I said. Kyle hung up quickly; like me, he was in business mode. I commanded Siri to make a few calls while I drove over the Chesapeake, watching the roiling storm clouds depart to my northeast. Within fifteen minutes of the tornado lifting, my MyRadar had granted FOX 5 DC permission to run the video, which Jason also embedded in a *Washington Post* article. I was pleased to see all my employers collaborating like a well-oiled machine.

Realizing I couldn't catch up to the storm, which was weakening anyway, I searched for an exit on the east side of the Bay so I could reverse direction. Something told me there'd be plenty of damage to cover. As I merged onto the highway westbound, my phone rang again. It was Matt Gaffney, FOX 5 DC's afternoon executive producer.

Apparently, FOX 5 DC *did* need me after all. He wired me into the studio for a live phoner. Seconds later I was on the air, the voices of Sue and Caitlin crackling out of the muffled phone speaker.

"Look at that!" gasped Sue, who hadn't seen my video before it appeared on-screen at 2:44 P.M.; Caitlin was equally taken aback.

"I had started driving northwards, and very quickly the funnel touched down, rapidly condensed . . ." I explained. I was driving and trying to follow the GPS, but that didn't stop me from guessing what was on-screen and spending a half hour doing live tornado coverage. "I can confirm it was on the ground for at least twelve minutes," I said, examining the time stamps of when I had first witnessed the cone funnel from the highway. "This was not your typical tropical tornado."

Eventually, I managed to exit the highway and pull into a dirt strip on the side of the roadway, where I fired up the MyRadar app, pulled up "superres velocity tilt 1," and broke down what I was seeing on air. *I am the tornado guy*, I thought victoriously.

After a half hour on air, Matt Gaffney called again: he wanted me to stick around and cover the damage.

"You guys have a crew for me?" I asked.

"Yeah, but we need to go live in forty minutes," he said. "Crew won't be there until 5:30. Can you Skype in?"

"Sure," I said. "Except I'm wearing gym shorts and a soaked T-shirt. Y'all have a FOX shirt or something for me?"

"It's fine," he replied. "We're just glad you're there." I smiled, even though he couldn't see me.

I thought back to minutes prior. Sue had mentioned damage on West Street, so I set my GPS for there. I tapped the TRAFFIC feature on Google Maps so I could hunt for road closures. That would help me find where the tornado had actually hit.

By 3:20 P.M., I was in the thick of it. The twister had crossed West Street, Annapolis's main artery, and by the looks of it, it was a strong one.

A Wendy's and Burger King had been heavily hit, with police reporting a gas leak at Sun and Earth Natural Foods. The roof was missing off an ethnic grocery store, and homes on neighboring Parole Street were largely destroyed. I estimated at least low-end EF2 damage.

A police line stretched across the road, with the sounds of chainsaws, walkie-talkies, and heavy machinery filling the air. Fire trucks were parked in the road next to me. I looked like a college student with an iPhone, but I was preparing to lead a special report in major market TV. The sun was once again shining. A coating of leaves formed a stucco dressing on the neighboring Exxon station. I could tell the vortex has swirled just to the south.

Gaffney sent me a Zoom link, and at 3:30 P.M. I was recounting my experiences on-air to the sincere shock of Rob and Angie, the afternoon anchors. I still couldn't believe what had happened—no TV meteorologist in the Washington market, or on the East Coast for that matter, had ever filmed a report in front of a tornado in their DMA as far as I could tell. I hoped that management was watching.

It occurred to me that I wouldn't make my 6:00 P.M. dinner plans. I was meeting Patrick, a recent Hopkins graduate and friend of mine who was a math teacher at a private high school in Wasington, D.C. Allen, and my other friend Sauhil, were busy.

"Hey . . . so . . . I got into a tornado," I typed at 4:18 P.M., squinting in the sunlight; the wet ground reflected a surplus of light. "Is late feasible?

"Damn are you swirling above the city as we speak?" Patrick replied, obviously convinced I was joking. I said nothing, and instead sent the video.

"OMFG," came the next message. "STAY SAFE."

I lead the 4:00 P.M., 5:00 P.M., and 6:00 P.M. newscasts from West Street; Rob, a photographer whom I had never met before, arrived around 5:30. I was shivering by then, still damp and coated in a rime of sweat and rainfall. Cold air was blowing in on the back side of the front.

He brought me a water and offered me a warm seat inside his vehicle. I graciously accepted.

Around 7:00 P.M. we ventured a block west to a torn-apart subdivision. Somehow the tornado had missed the Walter S. Mills-Parole Elementary School by barely a hundred feet. The dandelions on the school's lawn seemed to be carelessly engaged in a whimsical repartee.

Between television hits I filmed videos and reports for MyRadar, hopping on a livestream for them at 8:00 P.M. to discuss what had happened. The radio hits had come pouring in as well, and my videos had circulated around social media with more than a quarter million views. I delivered a live report on FOX Philadelphia at 9:30 P.M. By then, New York City was flooding.

As I had anticipated, extreme rainfall was occurring along the cold front. Newark International Airport picked up 3.24 inches in a single hour's time; New York's Central Park got 3.15 inches. Both readings obliterated one-hour records, and Newark saw their wettest day ever recorded with nearly eight inches of rain. At the time, I knew the Northeast was in the midst of a major disaster, but I had no idea that nearly four dozen people would perish in floodwaters that night.

A total of thirty-five tornadoes carved through the Deep South, mid-Atlantic, and New England during the passage of Hurricane Ida's tropical remnants, including an EF3 in Mullica Hill, New Jersey. That was the Garden State's strongest twister since 1990.

"Will 11:40–11:55 work?" I texted Patrick at 9:00 P.M. "I have to do the 10/11 news. Apologies in advance if I smell like rain and look like death."

I was drained from having been up since 1:00 A.M. and the emotional toll the day had taken, but adrenaline was keeping me going. Still, I needed food and an opportunity to decompress. We settled on meeting at Crystal City Sports Pub in Arlington, which was open until 2:00 A.M.

I could hardly remember who I was or what I had seen by the time the 11 o'clock news rolled around, but my on-air delivery was crisp and authoritative. Moments before I went live, an officer with the Annapolis Police Department knocked on the door of the vehicle in which Rob and I sat. He handed us a warm pizza. My eyes teared up from the thoughtful gesture.

My phone rang as I drove back to Washington at 11:30 P.M.; it was my parents on Cape Cod. Their phone had awoken them with a tornado warning; queuing up radar, I gave them an update.

I rolled into the Sports Pub at midnight; Patrick was already waiting.

"Sounds like you've had a busy day," he said with a smirk.

"And I still have to do DW at 2:00 and 4:00 A.M." I sighed as the waiter walked over. I ordered tater tots and a pizza. (If you've never had tater tots on pizza, you're missing out.)

I got a total of about two hours fifty minutes' worth of sleep that night, and was at FOX 5 DC the next morning at 8:30 A.M.; they dispatched me back to Annapolis during the afternoon hours for additional tornado coverage. I also made appearances on DW News, BBC News, and MSNBC. It seemed everyone wanted a word.

I had never been so tired. I was also the happiest I'd ever been. But nothing was better than the text that came at 5:39 P.M. that Friday from Paul.

"I want you to know I have bigger plans for you," he wrote. Kyle sent a similar message at 6:35 P.M.

"I wanted to send a quick note and let you know you did an amazing job this week," the text read. "Keep doing what you are doing and big things will come very soon. I'm glad you are on our team, Matthew."

Smiling, I glanced skyward at a vacant sky of cobalt blue.

"Thank you," I whispered.

Looking Up

Like most meaningful things, life is fleeting; it's ephemeral. The Earth has been around for 4.5 billion years. Humans get maybe a hundred at best.

Most people squander them, performing their societal sprints with tunnel vision as they race to complete artificial milestones; it's a real-life video game with levels, shortcuts, and pitfalls, but structured around the accumulation of all things tangible. Some spend years glued to their smartphones; others remain cloistered in the hollow comfort of routine, or become embroiled in conflict and trivialities for the sole purpose of self-occupation. It's an easy trap to fall into, but it's never too late to change course.

In my experience, the most memorable times in life are those that can't be planned, anticipated, or replicated. They're born from spur-of-the-moment decisions that often defy convention or tradition but often are the result of curiosity. They stem from seeking the hidden adventures wrapped in the everyday.

They're daring someone to join you on a last-minute flight, or cramming your mother into a double-decker bus with nothing but mini muffins and stale pizza. They're pondering the purpose of your existence while being baptized by the moon's shadow, or doing doughnuts in a U-Haul van in a snow-covered Alaska parking lot. They're catching a case of the giggles wandering the aisles of Walmart in search of late-night cake, or scarfing down leftover Waffle House at a rest area while listening to your favorite song. They're shielding your ears from the piercing crack of nearby crashing thunder.

Adventures can happen anywhere. Sometimes at airports, occasionally in the desert or a gas station, and, once in a while, in bingo halls. Adventures don't follow a schedule or adhere to a prescribed itinerary or time zone. They can peak at 2:00 A.M. amid a visitation by the northern lights, or at 2:00 P.M. as news breaks inside a television studio.

Adventures aren't always easy. In fact, the best ones usually aren't. They involve steep uphill climbs, heartsinking downward slides, and many sleepless nights. They might require sitting on a thirty-pound bag of onions for six hours in a haggard minivan missing a few seats. A good adventure can entail sprinting through a foreign bus station or replacing a fractured windshield. Usually, the experience is worth the hurdles; some adventures can take years.

Adventures can come when they're least expected. Several of the storm chases that left me with the greatest lasting impressions were those when I wasn't actively chasing. They took the form of mesmerizing sunsets or a thin bank of mist catching the sun during an episode of South Dakota hail fog. Sometimes a day's best adventure is spotting a pronghorn antelope roaming the prairie and not running it over with your truck. It's easy to miss the beautiful moments by not noticing them. Life is filled with everyday simple gifts.

The most important adventures are those that involve other people. They manifest in shared laughs, inside jokes, and instances of "you had to be there." Any moment can be made special by the right person or people. The key is finding those people and surrounding yourself with them. They're as rare as EF5 tornadoes. Once you have those people, you do everything to keep them around.

As I write this, I'm groggily flying from Chile's Atacama Desert to Patagonia, a reserve of fauna, volcanoes, ice mounds, and lagoons at the tip of South America. We just flew over the Perito Moreno Glacier. (That's another rule I live by—always take the window seat.) I'm running on fumes, having sacrificed sleep and remained up most of the night photographing the annual Geminid meteor shower from the famed Valley of the Moon. Meteor showers are good for the soul.

At the end of the day, I want a life of looking up. Of learning to appreciate and understand the beauty that life delivers; of making the most of those beautiful moments. I plan to die with an empty bank account and a full passport.

My unorthodox and, frankly, bizarre path has taken me down avenues most folks wouldn't ever set foot on, but I relish the uniqueness.

Years from now, I hope to have my own show, centered on getting off the beaten path and bringing viewers along for the ride—showing people things they might never have even imagined. I want to introduce the public to new cultures, teach them some science, and offer a taste of the truly incredible world that exists beyond the realm of what they know. I want to chase new horizons.

Ultimately, I hope to instill in people that, with the right people and a daring attitude, every day can be an adventure. Sometimes it all starts just by looking up.